RC

BUILDING TECHNO[LOGY] 3

CONSTRUCTION TECHNOLOGY AND MANAGEMENT

A series published in association with the Chartered Institute of Building.

This series covers every important aspect of construction. It is of particular relevance to the needs of students taking the CIOB Member Examinations, Parts 1 and 2, but is also suitable for degree courses, other professional examinations, and practitioners in building, architecture, surveying and related fields.

Project Evaluation and Development
Alexander Rougvie

Practical Building Law
Margaret Wilkie with Richard Howells

Building Technology (3 volumes)
Ian Chandler
Vol. 1 Site Organisation and Maintenance
Vol. 2 Performance
Vol. 3 Design, Production and Maintenance

The Economics of the Construction Industry
Geoffrey Briscoe

Construction Management (2 volumes)
Robert Newcombe, David Langford and Richard Fellows
Vol. 1 Organisation Systems
Vol. 2 Management Systems

Building Contract Administration and Practice
James Franks

Construction Tendering: Theory and Practice
Andrew Cook

BUILDING TECHNOLOGY 3

Design, Production and Maintenance

Ian Chandler

B.T. Batsford Ltd · *London*

in association with the Chartered Institute of Building

First published 1987

Reprinted 1989, 1992, 1994

Typeset by Progress Filmsetting Ltd
and printed and bound in Great Britain by
Redwood Books, Trowbridge
Published by B.T. Batsford Ltd
4 Fitzhardinge Street, London W1H 0AH

British Library Cataloguing in Publication Data

Chandler, Ian
Building technology.
3: Design, production and maintenance.——
(Construction Technology and Management)
1. Building
I. Title II. Series
690 TH145

ISBN 0-7134-5018-5

Contents

Preface

This, the third volume on building technology in this series, is concerned with some of the wider issues affecting the construction of buildings. The first two volumes deal with the application of scientific principles in establishing the performance criteria of buildings (wlth their effect on building deterioration).

The role and integration of services has been discussed, together with a comparison of the ways of building. On this foundation is introduced here the wider ramifications in the use of the technologies during the three major phases of a building's life, namely design, production and maintenance. Initially the factors that constitute a contextual framework will be discussed. These factors attempt to give the boundaries within which technological decisions in building are formulated. They are not deemed to be all that shapes and sways decisions of a technological nature, but they will designate areas within which issues can be focused. These factors must also be considered in the reading of the first two volumes: the three volumes are to be seen as a whole which the conceptual framework embraces.

The manner in which the issues associated with the framework are presented and discussed reflects my conception of building technology at a level equivalent to the final year of a BSc degree. There is little or no reference to building construction details. It is expected that these will be known and understood in relation to their specific design and construction criteria.

The purpose of this volume is to tease out those issues, concerns, influences and factors which contribute to the achievement of optimum technological solutions. The processes of analysis are explored and how they can be used in the decision making process. I try to show that the route to a technological solution is a complex matrix of disciplines, interests, attitudes, facts, materials and methods. In most cases I have not given answers, preferring to leave the student to answer the questions posed. The answers will be found in the practice of building technology. Most, if not all, buildings are unique and it is almost inevitable that today's solutions will not be tomorrow's remedies!

Society is subject to continuous change and development. Consequently, this book must be seen as presenting issues and concepts and techniques for analysis; it does not offer any prescriptive solutions. The reader will need to refer constantly to other reference material for in-depth explanations of techniques and practices.

The questions appearing at the end of each chapter cannot necessarily be answered by direct reference to the text. They are designed to direct attention to current practice in technology and to provide the opportunity to make use of experience. There are no absolute answers!

I wish to thank my colleagues at the Polytechnic of the South Bank, especially Adrian Cridge and Aldwyn Holley, who together with Alan Bleasby, are dedicated to the advancement of the discipline of building technology. These three volumes can, in some measure, be regarded as a culmination of their work in developing the teaching and learning and syllabus material for building technology.

Thanks must also be given to Professor Michael Romans for his extremely perceptive and constructive suggestions made during the development of the technology curriculum.

Part One
A CONCEPTUAL FRAMEWORK

1. Introduction

Ours is a complex society, not only in a technological sense but in regard to personal relationships. Individual lives are moulded by many factors not within direct or even indirect control. Although our society allows changes in policies, ideas, morals etc and an ability to voice opposition, any decision or course of action is nevertheless constrained in some way. Constraint may be covert, as with our upbringing or cultural heritage; or overt, in the form of legislation or established practice.

The decision to buy a house and become an owner-occupier is as much a cultural trait as a natural desire: the great majority of other industrial nations rate home ownership low on their scale of priorities. This is not to say that they do not demand a high standard of accommodation, but the desire to become actual owners is low. This social attitude influences the type of buildings constructed, the technology used and the expectations with regard to standards.

The intention of this chapter is to point to the factors that influence technological solutions for buildings and to provide some explanations. The factors discussed are also highly relevant to the material presented in *Building Technology 1* and *Building Technology 2* in this series.

Building technology cannot be isolated from the society in which it is practised and developed. To gain an in-depth understanding of technology solutions it is necessary to consider all those aspects which have come together (and the process by which it occurred). A relatively simple decision relating to the choice of a dpc in a brick wall is governed by many factors, as shown in Fig. 1.1. Because there is a wide range of choice in dpc's the final form must be suitable for the application. Any decision which ignores the realities of prevailing conditions can produce premature failure. A possibly apocryphal story about the Russian Orthodox Church on the eve of the Russian Revolution claims that the main item for debate was the choice of colour for its ceremonial robes. In other words, the clergy failed to respond to the major factors which were threatening its very existence. Similarly, a technological decision

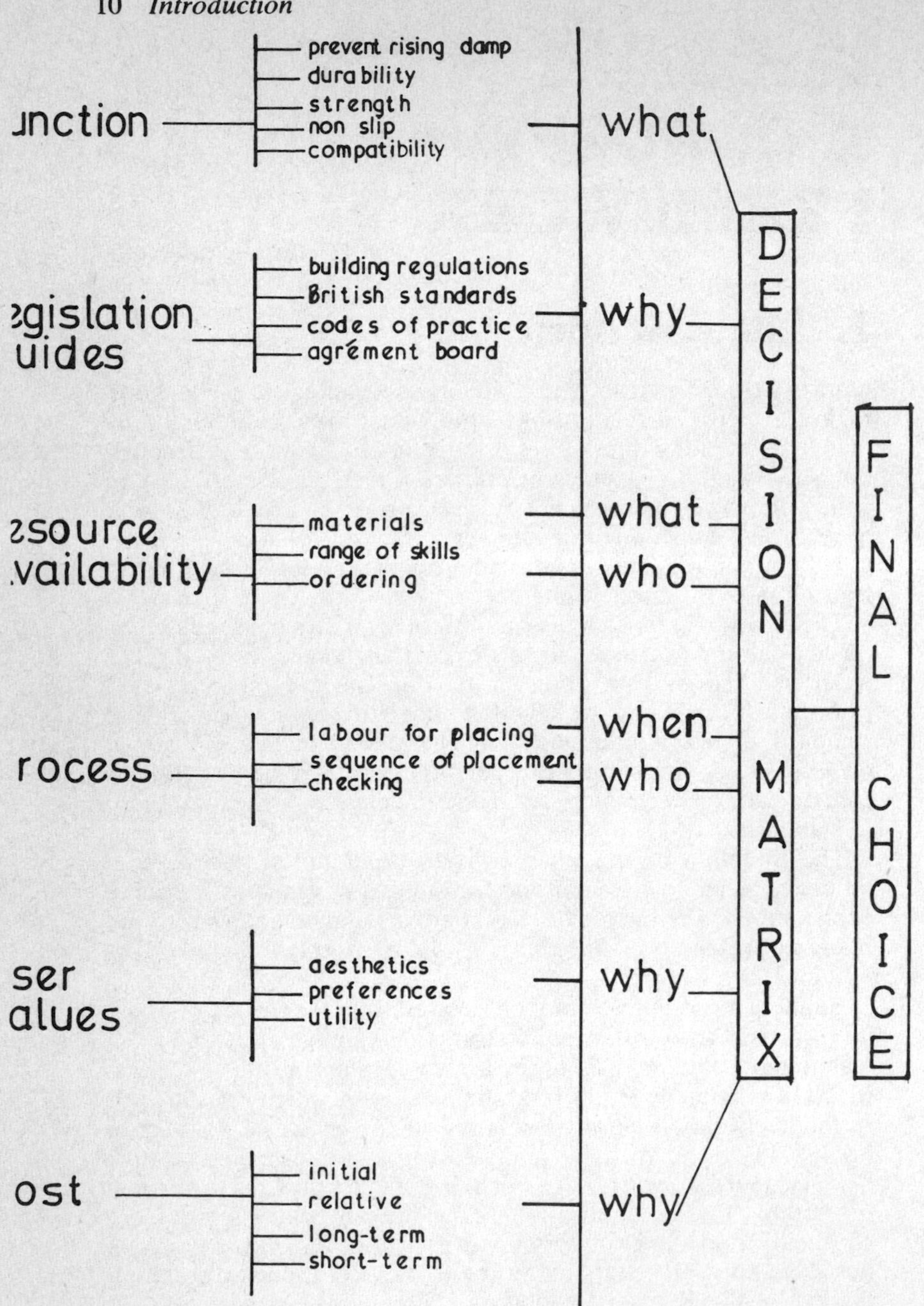

1.1 Factors affecting choice of damproof course

which ignores its context can at best perform inadequately and at worst fail completely.

In Fig. 1.1 it can be seen that the range of variables is such that questions need to be asked. Which questions are just as important as the answers obtained, for only the right questions will lead to our optimum solution. At any one time any of the factors listed may exert a strong influence over the final choice. It may come down to a bottom line of cost; or recent legislation affecting standards required; or the need to reduce the visual impact of a horizontal line across the face of the brick wall. Another individual faced with the same problem may select a very different solution.

Why? One cannot profess to give an answer to this as it implies an investigation of the rationale of each, followed by a detailed comparison. What is important is that those factors are defined which are considered in the framing of the solution. It is then up to the decision makers to make the selection, hopefully now with a greater awareness of the factors which affect the choice.

If there was but one technological solution to a given problem then the built environment would be stereotyped. That there is such variety shows that there are many solutions to the same problem. There are, of course, basic principles which are common to most solutions and these can be seen in particular construction details and building elements. For the purpose of this volume it is assumed that these basic techniques and the functional reasons for their adoption are understood.

Referring again to the dpc, the need for a dpc must be appreciated, together with the materials used, position and method of installation.

Building technology is concerned with the wider issues affecting the rationale of dpc's, not with the techniques of construction *per se*.

As science and technology advance to create new understanding, specialisms are created in order to deal with them. It is not possible for one person to understand all facets of a particular discipline.

As they develop, specialisms must be seen in relation to overall development and practices not only in their own field but also in allied fields. This can be illustrated by the jobs and functions within a builder's organisation (Fig. 1.2). In comparison is a typical pre-war firm operating with a manager, site managers, site clerks and a visiting quantity surveyor. The workload was two or three estates each comprising 100 to 200 houses. The quantity surveyor had responsibility for all sites, and all site functions were performed by the site manager and a clerk, with the able assistance of the trades foremen.

It may be argued that technology has necessitated the separation and creation of functions. An alternative viewpoint is that no real

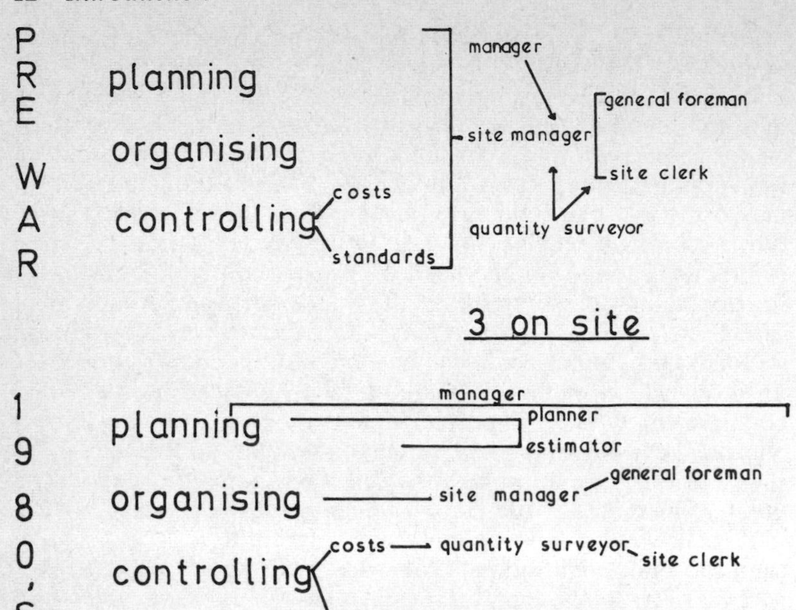

1.2 Site management personnel

advance in house construction technology has been made but that a shift in responsibility has occurred from the site craftsman to the administration. No doubt the answer lies somewhere between the two, but there is no doubting the inexorable trend towards a reduction in site activities. The increase in functional activities does mean that effective communication is now more difficult. The Tavistock Institute report *Interdependence and uncertainty* gives a valuable insight into this problem.

The pre-war manager had responsibility for planning, organising and controlling the site. Nowadays he may have the same degree of overall responsibility but the tasks will probably be undertaken by separate people, as shown in Fig. 1.3. Now there are six individuals making decisions, creating a multifaceted communication network.

The technological advances that have spawned specialisms have created the need to ensure that the technological co-ordinates are integrated effectively in the process and the product. The study of technology involves learning about environmental science, properties of materials, construction techniques, building services, etc. and the successful integration of these in the realisation of a construction project. Knowledge is of little use unless it can be used. The fitting together of the parts, in both physical and organisational terms, is the responsibility of the building technologist.

```
                      ┌─site manager ─< assistant
                      │                 assistant
                      ├─general foreman
contracts             ├─clerk
manager               └─site engineer

                      ┌─planner ─< precontract
                      │            contract
                      ├─quantity surveyor
                      ├─bonus surveyor
                      ├─buyer
                      └─plant manager

                      ┌─administration
                      ├─secretarial
                      └─computer
```

1.3 Management functions

If we are to make technological solutions in context it is necessary to identify and explain the constituents and range of that context. To make sense we have to label and allocate ideas and then to relocate them in the new situation. The divisions represented here are one group of labels upon which we can centre discussions. It is not intended to be prescriptive nor extensive. You will find other ideas and factors that could form a rationale.

This group of factors should be likened to the skin of an orange – they form the boundaries of the individual fruit of which the flesh and juice are the body of the technology and the pips are the lessons for the future. The orange was attached to the branch and the tree, which was grown in a climate or environment. The pip produced the tree which produced the fruit. By manipulating this cycle we can alter the characteristics of the fruit. Likewise by manipulating the inputs to the building process we can change the product under controlled conditions, but only if we know the scope of the inputs.

The contextual framework provides the means by which the skin's characteristics can be ascertained, together with the supporting arteries of the branch and tree.

The contextual framework is divided into eight sections:

Functional requirements
The need to ensure the building operates efficiently.

Safety aspects
Safety considerations in relation to the construction process and the building's users.

Economic, planning and legislative factors
The effect that costs, economic policies, planning policies and legislation have on the process and the product.

Resources available
What resources are available and how these may influence construction.

Design
The integration of design and building processes. How does the separation of roles affect technological decisions?

Technological change and development
New techniques, materials, components are introduced daily. How can these be effectively incorporated into the structure? What initiates change and how can this be controlled?

Society/technology interface
The influence and pressures of society and its expectations need to be considered when producing buildings in order to satisfy wants.

User values
Each of us will place a different value on an object; each contributor to the building has his own aims and objectives. These may conflict and their resolution has to produce a worthwhile product which meets the subsidiary aims of the participants to the process.

The above framework sets the scene for specific topics within the area of building technology. Each topic will draw upon the contextual framework to the extent that this affects its resolution. At any one time individual factors may have a greater or lesser influence.

Fig. 1.4 gives those factors which are bound by the society/technology interface. The process of building integrates and binds them together. The following chapters in this section will further explain each of the conceptual factors and show where they may influence the technology.

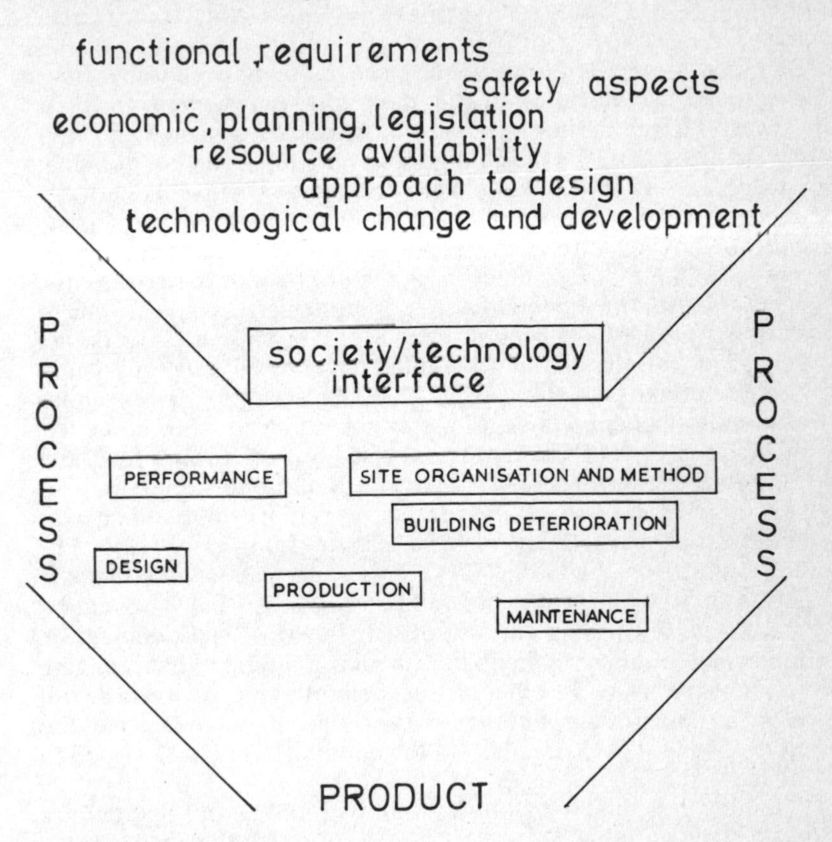

1.4 From contextual framework factors to product

2. Functional requirements

Buildings are created to serve the needs of man; even follies have given satisfaction to their builders and amusement to their viewers. However, the majority of buildings are designed and built to satisfy the need for the continuance and improvement of living, playing, resting and working. Buildings can be grouped generically under dwellings (houses, flats, bungalows, etc.); leisure (sports halls, swimming pools, cinemas, etc.); offices; factories; warehouses; shops etc. Certain criteria need to be met in servicing the needs in a particular building; it is difficult to convert one to another as most have been designed for a specific use. To design and build with the aim of satisfying two or more functions would be difficult and probably costly. Nevertheless there are examples of buildings where a change of use has occurred from necessity, such as the Covent Garden Market in London, from a fruit and vegetable market to shops, offices, cafes and public piazzas.

A move towards the concept of versatility in building was initiated by Alex Gordon with the phrase 'Long Life, Loose Fit, Low Energy' in the early 1970s. This concept demands that the design of a building take into full consideration that it should be constructed for a long and useful life, but that it also should be amenable to internal flexibility to accommodate the changing needs of the users. In other words it should be a shell which will allow a variety of solutions to the division of space and the provision of services. In addition the building has to be designed to minimise the use of energy.

Initially it is the use to which a building is to be put that defines its technology. It has been argued by Bowley that the need to solve a functional requirement is the main impetus behind construction innovation. An example of this is the need for tall towers for radio transmission. A structure was required which could be built with the minimum plant and machinery (it was inevitably sited in a remote location). Reinforced concrete towers provided an optimal solution, and the use of slip-forming techniques reduced the need for tall land-based cranes and allowed a continuous and speedy rate of construction. Without the need for tall towers in remote locations it is doubtful if slip-forming would have developed. Since the climate in the UK does not on the whole allow living and working in the open air, the majority of working time is spent within buildings. The buildings need to function irrespective of the environment and to provide protection from the wind, rain, snow, sleet and sun. It is the function of the buildings to ensure that its occupants are adequately protected.

Buildings cannot be constructed without taking climate into account. (Climate and the building are discussed in Vol. 1.) In other parts of the world the climate makes more exacting demands on building technology. Most buildings require internal warmth during the cooler seasonal cycles. Whatever the ambient temperature there are occasions when supplementary heating is required. This can be achieved by introducing an internal heat source through containing indigenous heat within the structure. In temperate and cold climates it is necessary to do both. A building must have the ability to accommodate both systems. The level of temperature required is determined by the type of human activity carried out within the building, which in turn forms the type of building itself. If hard physical activity is being pursued, e.g. playing squash, the body's need is for a low air temperature. Sedentary workers on the other hand will require a higher air temperature. The technological solutions must embrace the need for an auxiliary heat source as well as ensuring that heat is not lost nor excessively gained by the building fabric.

COMFORT

A building that works effectively is comfortable. The human being can relate to and enjoy the artificially constructed environment so long as it does not upset physiological and psychological balances. Comfort is more than just warmth. Le Corbusier starkly described houses as machines for living in. In his interpretation of this concept he designed houses which were purely functional. The building was capable of being designed using engineering criteria related to the ergonomic needs of people. He also implied that as a machine it should be efficient and effective. Every component, element, texture, surface and fitting should satisfy just one prime question, 'Does it function adequately?'. In the practice of this concept Le Corbusier's buildings (and those of his 'disciples') were criticised as being too clinical and 'cold'. Some internal environments were emotionally unfulfilling. People like to make a house a home: to put a mark upon the space in order to link identities and gain security.

Other building types seek another form of comfort. Workers in offices with little access to natural light have been found to suffer from depression and lack of work motivation. To have some natural light will enhance the feeling of comfort. But for some the view of the great outdoors will provoke a desire to flee from the workplace and enjoy the leisure attractions! A building needs to pander to people but not distract them from their activities within it.

SAFETY

A functional requirement of a building is that it must be safe to its users and to third parties. The structure must have stability and strength. It must resist climatic forces. It should not be adversely affected by fire, allowing enough time for the occupants to gain safety via adequate means of escape. A collapse of one section should not provoke progressive collapse of any other part. Volume 1 has dealt with some aspects of fire technology and we shall reconsider it amongst wider issues later in this chapter.

DURABILITY

Generally buildings are expected to have a long life. Indeed, nearly 50% of English housing stock is pre-1939. It is also expected that the maintenance requirement to buildings should not be excessive, although it is recognised that certain materials do have a limited life. Durability can be compromised by a number of factors, such as poor cheap materials, inadequate design, poor initial workmanship, lack of maintenance etc.

There is an inevitable compromise to be made, in the choice of materials and details, between the desire to ensure longevity and the cost of the design solution. As advances are made in the production of new materials and in the understanding of their behaviour, the life of buildings and their constituent parts can be more accurately forecast. This will help in the formulation of solutions to satisfy user demands. In some cases a limited life of only a few years may be wanted and the structure and its materials need not be of a great durability. It should in all cases be possible to select those that will serve the optimum term.

The question of durability is becoming of increasing importance. As previously mentioned it has been incorporated into a tripartite alliance within the concept of 'Long Life, Loose Fit and Low Energy'. Further developments are being made to bring a wider awareness of the need to design and build with durability in mind. Such techniques as value analysis and terotechnology are means by which the long-term costs of buildings can be ascertained. (This will be discussed in the section on Design.)

At the tertiary level of the functional considerations, those relating to value are shown in Fig. 1.5. This value can be measured in monetary or social terms.

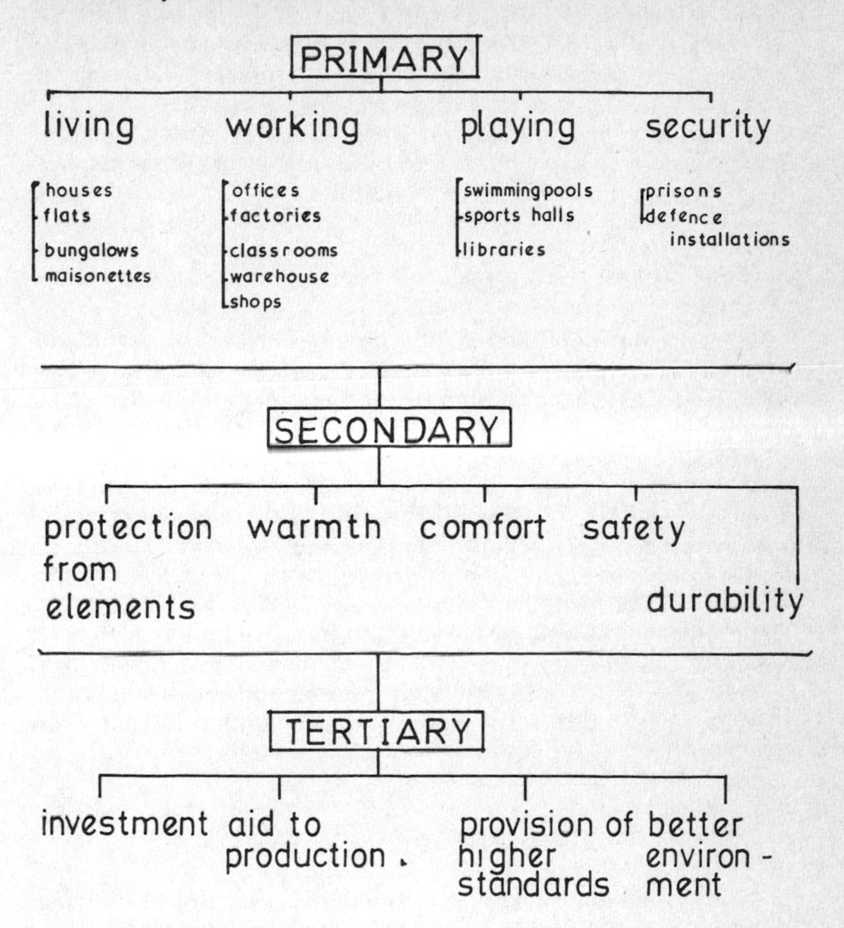

1.5 Hierarchy of functions

INVESTMENT

Any building, if built to a good standard, can be regarded as an investment. The form in which this benefit is realised may well influence the technology used. For example, a building may be regarded in purely monetary terms by private developers who see their buildings as accruing assets. In order to maximise the investment the technology is subservient to the balance sheet. On the other hand a private house purchaser, whilst seeking a sound investment, is not motivated solely by the future possible value, but is concerned that the house meets his demands.

Buildings may be used as collateral in the financing of other buildings or businesses. Loans can be raised against the value of the building or, if mortgaged, against the increase in value over the outstanding mortgage; there has been a tendency for the great majority of buildings to appreciate in value over time.

The building stock is a major contributor to the nation's wealth. The nett capital stock of buildings and dwellings in 1983 in the UK was accounted for about 70% of the total.

AID TO PRODUCTION

A NEDO report published in 1978 put forward strong arguments for major investment in the construction of new factories. (Up to 1985 there has been no significant increase in this sector of construction.) It was said that by settling existing and new industries in new factories, greater efficiency would be achieved. Many industrial practices could not be readily improved owing to inadequate buildings, and modernisation of these would be costly and involve temporary shut-down. As the UK relies heavily on exporting goods there is a strong case for the provision of new, purpose-built factories.

A particular example of the need to improve buildings to cope with new practices is in the sphere of information technology. Computers require controlled environments and complex services. It is extremely difficult to adapt old buildings to suit their demands.

PROVISION OF HIGHER STANDARDS

Society is demanding ever higher standards in all aspects of living. Increases in real wages in recent years have enabled standards to be enhanced and this has had its effect on building technology. The incorporation of a bathroom en suite with a bedroom to a house, central heating, and perhaps double glazing are typical examples.

IMPROVED ENVIRONMENT

It is not just the architectural style of a building that makes an impact on the environment. Poor constructional detailing can manifest itself in the staining of the facade; buildings which cannot cope with manufacturing waste may cause atmospheric pollution.

In the hierarchy of functions shown on Fig. 1.5 it could be that the primary level should consist of those functions relating to human needs. After all that is what buildings are for. But here we are concerned with the technology of the building, not the need for the building. If this is the aim, then, especially as we live in an age which can satisfy the basic needs for shelter, we must realise that we have many choices. It is accepted that there are many examples of building types and that we do not have to reconsider the provision of shelter as a function. Our level of ability and knowledge transcends the provision of simple shelters for any activity. In an industrial society we are concerned with satisfying the requirements of diverse human activities and we are now developing the technologies to sustain these.

3. Safety

There are two aspects of safety that need to be considered. Namely, that which effects the safety of operatives, the public and the structure during construction, and that which affects the users of the building and third parties during its life.

During construction the builder needs to be concerned about safe working conditions for the work force. Unfortunately, a general disregard for safety in the past has necessitated the enactment of legislation to enforce good habits. Even with this legislation, and its associated penalties, indifference to safety is still endemic. Deaths and serious injuries actually increased during the Site Safe campaign in 1983 to equal levels recorded in 1980. These increases have occurred with a greatly decreased work force and point to a serious problem which needs further investigation. The industry's safety record poses questions about the incidence of sub-contracting, self-employment, competition and general attitudes towards safety.

The need to enforce adequate safety precautions has a profound effect on construction methods and sequence of operations. For example the need to erect scaffolding to the correct standard will prevent a continuation of work at that particular place; alterations to the scaffold will cause the access flatforms to be temporarily withdrawn from use.

Care must be taken by the Contractor to prevent injury to people in the proximity of the site. This could involve the use of protective netting, hoarding, fencing, fans, walkways etc. and their maintenance in good repair. The use of netting as a shroud around a partially built structure may hinder the installations of external claddings, windows etc. Alternative methods will be required to swing the components into place via preformed openings or special access routes.

Questions – such as: can the materials hoist be effectively used? will the components' sizes be restricted by access problems? – need to be asked in order to achieve optimum planning of safety measures.

The structure of the building has to remain safe during

construction and this may necessitate the use of temporary works to support the incomplete parts. The design and construction of temporary works can be just as complex as the building itself. Examples are steel frames as support to building facades whilst new work is carried out behind, and formwork support to slabs and slipform walls.

There have been many instances of building failures (see Stephen S. Ross, *Construction Disasters*, 1984) before or soon after completion. The strength of a framed structure is dependent upon the sum of its constituent members. If some are left out (owing to lack of time at the end of the day, wrong ordering or negligence) then the stability of the structure will be affected. Any excess or eccentric loading may cause a collapse. Even a simple brick wall may require support whilst the mortar is curing, or its height and/or length may be limited. As buildings become more complex and make demands on narrower tolerances and pared safety margins then the need for sound construction methods increases.

A number of accidents involving falsework led to an investigative report (*The Bragg Report on Falsework Design*, 1975) which has resulted in the consideration of supportive works design at an early stage in a project. Falsework must now be properly designed with the use of sound engineering theory and practice. A consideration of these aspects could influence the building's subsequent technology to a great extent.

The safety of a building throughout its life must be geared to its users and third parties. Decisions will have to be made relating to the degree of risk which is acceptable. For example, it would be possible to construct a building having a high resistance of fire, together with superlative means of escape and fire-fighting equipment. But this would inevitably cost a great deal of money. A balance needs to be struck between cost and risk.

The main structure element needs to be absolutely safe and as technology and science advance there is greater confidence about the behaviour of structures. This is translated into optimum design solutions to suit particular buildings. Unfortunately, in the past tragic accidents have highlighted the inadequacies of some of the systems adopted. The Ronan Point disaster in 1968, where an internal gas explosion blew out the external walls of a flat in a high rise building, is one such example. The system was erected on the principle that each wall and floor depended for its support on the one below. When one was blown out by the explosion those above were deprived of support and so collapsed. Now specific clauses have been placed in the Building Regulations to make designers aware of the need to ensure that if for any reason one part of the structure fails it will not lead to progressive collapse.

Every element and component incorporated in the building

needs to be safe. Recently the dangers to health associated with the use of asbestos-based products in buildings have been recognised. This has resulted in extensive removal and substitution. Alternative materials and products have now to be specified. This may have repercussions on the choice of the other finishes and details. For example, asbestos is a good fire-resisting material and is used as cladding to structural elements to give protection. Where found these have to be removed and some other means of protection applied. In the consideration of alternatives the existing service runs will have to be taken into account if they were also protected by the asbestos sheeting. Will the new material match the existing finishes? Can access be gained where required?

The above example shows that these contextual framework factors are not solely related to new building work. They are equally applicable to maintenance, conservation, conversion and rehabilitation.

Safety is not just an objective consideration, it is an emotive issue. The pressure for improvement is constant but changing in its demands as new dangers are discovered. New solutions will then be required from the building technologist.

4. Economic, Planning and Legislative Factors

These three are grouped together as they can be identified with the processes of government at both national and local level. Government policy and practice have a profound influence on the construction industry and in particular on building technology. An example is seen in the history of multi-storey structures for housing. In the late 1950s and early 60s a major political issue was the provision of housing for everyone, especially in the public sector. The two major political parties played electoral promissory leapfrog in quantifying the number of homes they would build when in office. Figures such as 300,000 dwellings a year were commonly banded about. These pre-election promises then had to be fulfilled. Unfortunately, the capacity of the industry to provide this number of dwellings by traditional methods did not exist. The solution then found was to build with untried systems. Planning requirements were eased and regulations set up to ensure that local government could provide the land and obtain the money. In addition government provided advice on suitable systems and encouraged their use. The reasons put forward in their favour were that:

(a) system building was quicker than traditional methods;
(b) dwellings could be constructed relatively cheaply;
(c) high levels of quality control could be achieved if units and components were manufactured under factory conditions;
(d) less land would be used if buildings were high rather than low;
(e) traditional craft skills would not be required, therefore easing the demand for them;
(f) the systems offered methods of building high.

It was the last point which had the greatest influence over the technology employed. With hindsight it can be appreciated that there were many reasons why the expectations were not met. In later chapters these reasons will be explored. In terms of economics, planning and legislation the creation of high rise flats was a direct response to government policies.

ECONOMIC

Many will argue that the prime influence on any technological solution is monetary – how much will it cost? There is little doubt that this factor is never far from the surface in the selection analysis. In any comparison there is a tendency to quantify the outcome – in other words, one aims for the optimum solution for the least money.

There are many ways in which economics can affect building technology, ranging from short-term and long-term investments to accounting and taxation procedures.

In coming to an investment decision, the life of a building and its suitability for user function should be considered in calculating the monetary return. Can the capital outlay be recovered over the life span of the building? Is short life technology used to keep down initial costs?

Monetary inducements, such as the giving of grants, may favour one type of development over another, with its repercussion on the technology employed. The use of a grant may enable a more expensive and longer lasting solution to be adopted.

Tax inducements can influence decisions. For example, capital costs have to be borne from profits or borrowing, but maintenance and running costs can be offset against future profit and therefore save tax. In this case it could make economic sense to build cheaply initially (implying low standards) and spend money as and when necessary in maintenance and remedial work.

The rate of interest for borrowing money is based on that recommended by the Bank of England acting for the government. It is common to borrow money not only to finance the development but also the builder during construction. Any fluctuation in interest rates will affect the amount of 'real' money available. A decrease will allow more to be spent whilst an increase could restrict the actual amount spent on the building.

Certain types of development can be encouraged by the easing of rates/rent for users. This might encourage more buildings of that type. An example is the setting up of enterprise zones to promote new businesses in an area, leading in turn to a demand for more buildings.

Another example is the setting up of workshops for small firms, where purpose-built accommodation is provided and existing property such as warehouses is converted.

Inflation can be a double-edged sword when related to building development. If inflation is high the costs of construction are high, which may lead to the use of alternative, less viable, technologies. On the other hand, in times of high inflation the value of buildings has tended to appreciate at the same rate. A corollary to this is the disturbing fact that building costs have risen at a far higher rate than general costs.

An analysis of economics as it relates to the construction industry is to be found in Hillebrandts book *Analysis of the British Building Industry*, 1984.

PLANNING

Planning is carried out for the benefit of the community as a whole, although in practice it may restrict groups or individuals in the achievement of their particular aims. Local authorities have produced plans which show the type of development that is allowed in certain geographical areas.

These may for example be designated as industrial, housing or commercial zones. In these areas it is envisaged that only that type of building will be erected, but there may be exceptions.

House builders have long argued that more land should be made available by local authorities. They claim that a shortage of land artificially raises land prices, which increases the costs for builder and purchaser. If land was more freely available then the price would be more stable.

The height, size, shape, colour and materials of a building can be controlled by the local authority planning committee and its officers. Decisions will be based on the need to blend the new proposals into the existing environment, both in the appearance and the use of the building. Factories will not normally be allowed in a housing estate or vice versa. Even the change of use of an existing building will require planning permission.

Upon the outcome of the application will depend the technology used. A proposal to convert a house into offices would incur the provision of access corridors, specialist services, appropriate means of escape in case of fire, etc. A refusal might lead to an alternative proposal, such as flats, which would entail the separation of the units, construction of bathrooms and kitchens, fire protection between floors.

Each solution would demand different technologies.

Planning decisions can also be seen as subjective. Many an argument has centred on the appearance of a building. In 1984 proposals were published for the extension to the National Gallery in Trafalgar Square in London, following an architectural competition. Prince Charles gained national headlines with his opinion, expressed at an RIBA dinner, that the proposals were like a carbuncle on the face of an old friend. Later the local planning authority declined to approve the proposals, on the issue raised by the prince, saying that a tower featured on the plan was not in keeping with the other buildings in the square. Whether the prince was influential in this matter is not known. Obviously someone had liked the tower otherwise it would not have appeared on the drawings. Beauty is in the eye of the beholder!

Another way in which planning can influence building technology is through the length of the approval process. At present, to gain approval quickly it is advisable to produce schemes which are not innovatory. Thus, it could be argued that out-of-the-ordinary, avant-garde or simply new technologies may cause disruption in the planning process. This may lead to increased costs owing to delays or may put off any future attempts at innovation.

A further example of the interplay between economics and planning is the boom in the provision of hotel beds in the early 1970s. The tourist authorities put forward strong arguments to show that the number of beds available was inadequate to meet demand, and that an increase in the number of tourists would be of benefit to the nation, especially in attracting overseas visitors with foreign currency to spend. The government was persuaded by these arguments and consequently gave powers to local authorities to allow the provision of more bedrooms. They also made available money in the form of grants payable to each new bed created. New hotels were erected and existing ones extended and improved; conversions were allowed. As the demand was there the hoteliers were able to recoup their outlays, more tourists were attracted to these shores and it was seen as a benefit to all parties.

LEGISLATION

Legislation that directly influences building technology centres on the Building Regulations. There are many other Acts which bear heavily but these will not be listed here since what is at issue is their general influence not their content or detail. The latter are best applied to particular construction projects when the occasion demands.

A fundamental question is whether or not legislation such as the building regulations is restrictive upon technology. For example, it has been said that the 'deemed to satisfy' clauses in the old regulations encouraged literal solutions to that requirement. By providing solutions they fostered a tendency to incorporate them in the building design and not think problems through.

If different proposals are submitted to the authorities then the designer has to justify them fully with calculations or examples. The contrary argument is that if specific details were not provided a wide range of solutions of varying quality would be offered by designers: each would have to be considered on its merits and monitored to ensure that the performance was as proposed. Also, many applications are made by individuals with little knowledge of construction and these clauses provide them with an interpretation which is acceptable. The basic purpose of the regulations is to ensure that agreed standards are met.

A case in point is the concern over the use of energy in buildings. Since 1966 the Building Regulations have been amended a number of times to reflect the national concern to reduce heat loss through the external fabric of buildings. This trend is likely to continue, especially as material manufacturers can now produce units with low levels of thermal conductivity. If the 'u' value is further reduced and materials alone cannot meet the requirements then the actual construction of the fabric will need to be revised, say by increasing the overall thickness.

A study of the new Building Regulations, operative since November 1985, shows a marked change of direction in that the legal regulation is separated from the offered technical solution. This means that the Regulations can, in the main, remain constant, whilst developments in materials, products and processes can be used to satisfy the legal regulation.

Legislation not directly oriented towards the construction industry may cause ripples that affect technology. For example, import controls could preclude the use of a common component, thereby requiring an alternative 'home produced' item to be used. This leads us to the next contextual factor, resources.

5. Resource availability

The basic resources for construction are land, labour, materials, plant and energy. Integration of these resources is an overriding function.

Each will be considered in turn, before their integration is discussed.

LAND

Land is a finite commodity but sufficient is still available to provide space for the needs of the built environment. As pointed out in the previous section the zoning and geographic location of land is important in assessing its worth for building. A relative scarcity of land in town centres will promote construction which utilises minimal structural framework and element sizes and maximises overall size in terms of height and volume. This in its turn produces problems for the builder. Space is limited for the movement and storage of plant and materials. The proximity to other buildings necessitates extreme care in maintaining their stability and condition. Major works such as underpinning may need to be carried out.

The problems of town sites are manifest, especially if there was a building previously on the site. The new foundations will have to take account of poor and cluttered ground conditions.

If land space is not at a premium, then the shape, size and technology of the building may be very different when compared to a town centre building. Nevertheless, whatever the available space, the technology must fit into the environment.

MATERIALS

If for any reason a material becomes unavailable, even in the short term, its lack may have a strong influence on the technology employed. A new material may also radically affect the technology by either superseding an existing material or allowing a new component or element to be used. A recent example of the former arose from the world shortage of copper. Pipework for water systems was the main user. A substitute material was required and the closest in performance characteristics was stainless steel. This was more expensive, harder to bend and required different jointing systems. These factors made its use more complicated and greater attention had to be paid to its limitations during the design stage. As soon as copper became available again it was therefore readopted for pipework.

An example of a material that arose from development is plastic. Now most sanitary waste pipes are in plastic. The lengths are butt-jointed with simple sealing sleeves, bends are glued on and lightweight fixing brackets keep the cost down. Whole sections can be easily prefabricated and fixed into position with the minimum of tools and skills.

The development of bentonite slurry or mud is a further illustration of a material directly influencing the technology. Without this thixotropic material it would be very difficult to construct deep basements in town centres. Its use in diaphragm walls has made it possible to create the earthwork support before digging out the basement. This also provides support to the surrounding buildings and ensures safer working conditions.

LABOUR

In this instance we shall consider only that labour which is directly employed on the site for production processes. Later we shall look at some of the issues relating to the whole range of skills for the construction of buildings.

On site (or in the factory production of components) there is a continuing tendency to reduce the level of skills required for the erection of new work. There is, however, still a need for a wide range of skills in the maintenance, refurbishment and conservation of buildings.

There is much discussion about the scarcity or otherwise of skilled people available to the industry. *Construction into the early 1980s*, published by the Building and Civil Engineering Economic Development Councils in 1976, attempted to provide some guidance on the demands that economic activity in the construction industry would make on labour and materials. This was mainly concerned with the overall effects when changes in demand took place, but it did identify trades or skills which would be particularly affected by changes. For example, those skills and materials exclusive to the construction industry which were most sensitive to changes in demand were: bricklayers (and bricks);

plasterers (and plaster and plasterboard); electricians; heating and ventilating tradesmen (used more in certain types of work than others); specialist operative skills (such as those employed in heavy civil engineering).

Within this overall demand pattern is another which can be attributed to the industry itself. Are there too few trainees entering the trades because builders do not provide training opportunities, or are new methods of production being developed in order to reduce the numbers of skilled personnel? Where the latter is the case it is being done under the banner of higher productivity and lower labour unit costs. Hillebrandt (*Analysis of the British Building Industry*, 1984) has shown that there is a general inadequacy in the provision of training. This is seen at its worst when an upsurge in demand is not fully catered for by the numbers and levels of skills. During recession training places are reduced – and since it takes at least three years to produce a satisfactory level of competence for most trades, there is a time lag between supply and demand. By the time a person is fully effective a dip in activity may have occurred, leading to under-utilisation of trained personnel.

One example of a change in skills required is in plastering. Now it is possible to spray a plaster finish on to walls using a machine. The rate of work is speeded up, provision has to be made for an energy source and the plant has to be moved around. The nature of the operation has changed and with it the technology and skills needed for wall finishes.

PLANT

Possibly the most rapid advances are being made in the development of plant suitable for the construction of buildings, in parallel with the sophistication and need for plant to service buildings. Power tools proliferate; materials are moved by fork lift truck; large heavy permanent service equipment can be installed by high capacity mobile cranes; small bulk material machines can operate within the building. So the list goes on!

The process and the product are becoming highly mechanised. New plant will affect the process of construction. New internal environment systems will affect the performance of the building, which in its turn will demand the technology to ensure its efficient housing. Perhaps Le Corbusier's statement that the house is a machine for living is becoming universally true.

ENERGY

In the design of buildings the level of consciousness regarding the use of energy has risen considerably since the early 1970s. Major research and development work is being carried out into ways of improving the performance of buildings as functioning environments. External elements are heavily insulated, heating systems are geared to use as little fuel as possible, and control systems prevent the unnecessary production of heat and air changes. No efficient and effective design will now ignore the running costs of a building's life. Mention has already been made of the national concern for energy consumption and its realisation in higher 'u' values for the external elements of a building. Present UK government policy regarding energy conservation is based on pricing energy at a high level. Compared to other industrialised nations the UK offers little in the way of positive aids to the conservation of energy, whether by adequate grants to improve existing standards or by promoting new ideas in the functional aspects.

Whilst there has been concern to reduce the overall consumption of energy, there is a tendency to adopt production processes which use more energy than in the past. For example, power tools and plant and machinery for almost all construction operations have increased the consumption of energy in two ways, firstly in the actual manufacture of the piece of equipment and secondly in making the machine work. There has been little research into the effects of a scarcity of energy on the plant used for the construction. Similarly there has been little energy auditing carried out on the manufacture of materials used in building. It is ironic, in a period of high unemployment, that techniques continue to be developed which are designed to reduce the amount of physical labour, thereby reducing the number of jobs and at the same time are increasing the use of energy. At the level of the individual firm this might make some economic sense but at a national level it is debatable. This point, relating to the use of manpower in the construction industry, is clearly argued by Hillebrandt.

INTEGRATION

Underlying the basic provision of the resources is the human activity required to ensure that these are brought together in the achievement of the building itself. It goes without saying that this integration should be carried out efficiently and effectively; this is the responsibility of the whole of the industry.

A favoured term to describe the co-operating functions is the 'building team'. As in any team individuals play their parts towards the achievement of one major goal. In building, the objective is the design and construction of a functional structure meeting performance and cost requirements. In some cases this objective is lost sight of in conflict between parties to the process.

The effectiveness or otherwise of this integration can have a significant impact on the technology of the building. Unclear specifications, poor drawings, inadequate information, muddled communications and careless quality control can directly affect the technology. In addition a general apathy towards the project can lead to inefficiencies in the process. Research carried out by the BRE on drawings has shown that the presentation of information could be greatly improved. Another report shows that poor communications between the building team can also have major repercussions on the technology. Therefore, the management of the project needs to be structured to suit the building technology. The people involved should be engaged with specific responsibilities relating to the technological demands. A traditional house construction will require a certain management structure, whilst an in situ concrete frame with precast units as floors and cladding may demand a very different structure.

An illustration of how the availability of materials can affect the mode of technology arises if one compares the structural frames used in the USA and the UK. In the USA the predominant material is steel: in the UK it is in situ concrete. The reasons for this can be traced back to the mid 1800s and to two main factors, material and knowledge. The raw materials for steel production were available in abundance in the USA. Large steelmaking centres grew up and the technology was developed. Commercial success fostered research. Universities became interested and with industrial backing created professorships and academic departments furthering research, not just in steel as a material, but in its application to building. The design of buildings in steel became a recognised discipline. This in turn attracted students and a growing body of knowledge came into being. This knowledge was fed back into the manufacturing and designing processes and thereby the science of steel-framed structures was further improved.

In the UK the desire to build tall buildings was not as strong as in the USA. Nonetheless there was a wealth of knowledge linked to the construction of frames in cast iron but this was not developed. For a comprehensive history of the use of prefabricated buildings, including cast iron, see G. Herbert, *Pioneers of Prefabrication*, 1978. The use of in situ concrete began to be common at the turn of the nineteenth century. In the UK there was an abundance of sand and gravel and the materials to make cement. This meant that the materials were relatively cheap.

Knowledge and expertise developed with the increasing use of concrete and eventually universities realised this and endowed professorships in the discipline, albeit a considerable time after the USA. Steel manufacture in the UK was not directed to the mass production of structural sections, so that it could not support strong academic research. Students studying design in UK educational establishments therefore concentrated on concrete technology. Recently a head count in a major design department of a nationalised industry showed that only two of the forty people were capable of designing steel-framed structures with any degree of confidence. However, this imbalance has been changing in the UK since the beginning of the 1980s (see *Building*, 16 November 1984).

The main reason for the increasing use of steel as a structural frame is primarily its relative cheapness. With a world slump and consequent over-production of steel, the price has fallen relative to other materials. This means that it is now economic to build relatively low rise structures in steel, despite the extra work involved in providing fire protection cover. If this trend continued there will be other repercussions in its wake. The other building elements will need to be related to a steel structure, different flooring structures could be used, also different claddings. Different skills will be required on the site as well as in the design office.

6. Approach to Design

A simplistic approach to the design of buildings could assume that each user requires a net area, that the accumulation of the areas produces a floor space and that walls, floors and intermediate floors in basic shapes will suffice. In reality the design process is much more complex; floor area is just one factor.

Another way to approach design is to concentrate solely on external appearance, creating a shape and facade which looks good and then trying to fit the functional requirements within. A further starting point might be the provision of certain rooms or accommodation, as for example three bedrooms and a kitchen-dining area in a house.

What is now beginning to exert a major influence over the approach to design is the need to ensure that the building performs efficiently. An analogy can be found in the design and choice of cars. Nowadays, a major consideration is not only the initial cost but also the running costs, both in regard to fuel and in maintenance and repairs. Advertising texts emphasise fuel economy and the low frequency of services – the fewer the better. We expect clean, efficient engines and areodynamic body lines. In addition, the so-called luxury items, such as wipers to headlights, fully adjustable seating, central locking etc., are becoming standard fitments. The performance characteristics of cars are compared one to another within the size and price range. This attitude and perception is now being adopted in the choice of buildings. Questions such as: How long will it last? What will be the maintenance costs? How frequently will maintenance be required? What are the heating and lighting costs? Will it be flexible to meet changing needs? – all these are now an essential ingredient of the design process.

In the UK this overall concern has been incorporated in the concept of terotechnology. Terotechnology is concerned with the specification, design, provision, installation, maintenance, modification and replacement of plant equipment, buildings and structures, and with feedback information on performance, design and costs. It can be defined as a combination of engineering, management, financial and other practices applied to physical assets, in pursuit of economic life cycle costs. The terotechnology sub-systems are shown in Fig. 1.6. It is a concept which helps to bridge the gap between the ultimate user of the building and the design and construction team. One of the processes should be a dialogue between the providers and the users of buildings and building products.

Fig. 1.6 also lists the other basic approaches to design. Those of space, function (as discussed in the first contextual framework factor), accommodation and appearance have been dealt with. Now we will address the question of performance.

In some countries, notably Sweden, the basis of building design is performance. Factors such as environmental climatic conditions are identified and internal comfort requirements are laid down to counter these. In so doing there are other criteria which act as constraints, such as the need to conserve energy. These criteria are quantified wherever possible. A total energy use is given for a building, internal temperature requirements are set and the

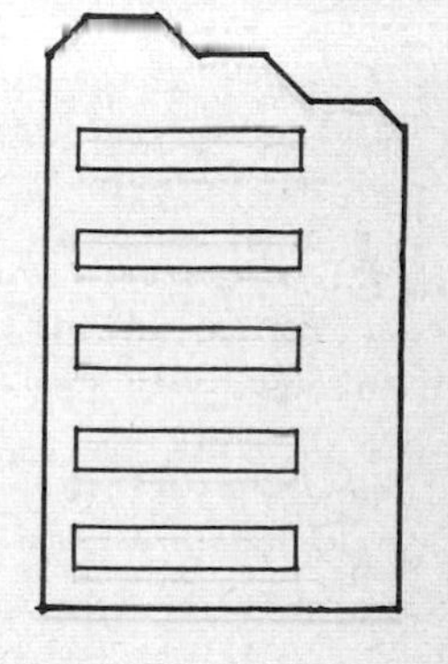

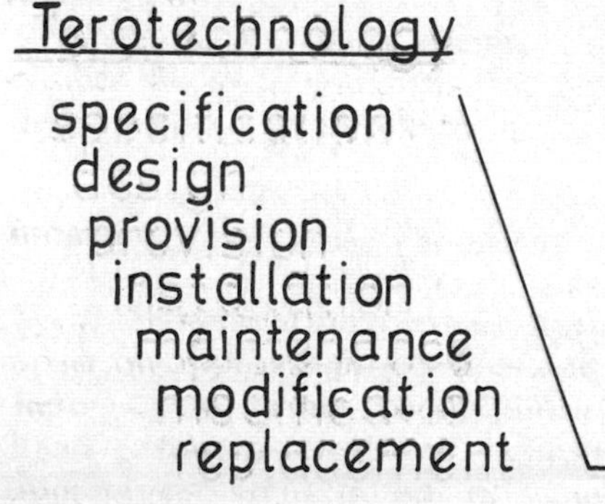

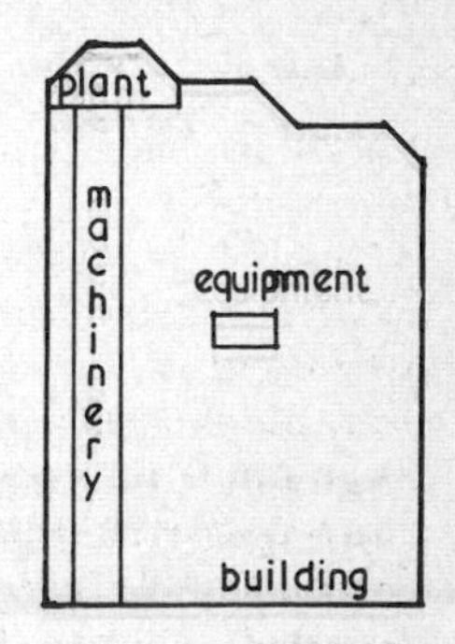

1.6 Approaches to design

technology has to provide the means by which these can be realised. No deemed-to-satisfy examples are provided. Each building is seen afresh and a technology produced to solve its particular problems. Indirectly, this provides more scope for the introduction of new materials and methods of construction. A set of norms is established which can be easily changed to meet new circumstances.

The criteria and norms for the preparation of a performance specification as a basis for design are set out in the BRE publication *The Performance of Building Components*.

In adopting the performance approach a rearrangement of the relationships between the parties in construction is necessary. It is unrealistic to expect one person or group within one discipline, to be fully cognizant of the choices available. In one area alone, building services, growth and development has been phenomenal. Neither the architect nor builder can expect to be able to accumulate the high levels of expertise now found in some areas of building technology. In the laying down of performance standards the responsibility for their achievement must rest with those that are providing that material or component. The specification details what is required: it is up to the builder, in conjunction with the specialist suppliers and manufacturers, to produce an item which will satisfy those performance requirements. For example, the client via the architect may specify that the windows will have to give a sound reduction of 20db; enjoy a maintenance free period of ten years; be openable and cleanable from the inside; give enough natural lighting for a specified distance into the room; perform to set criteria in regard to weatherproofing, joint specification etc. No proprietary window is specified or shown on the initial drawings. Prospective suppliers will be expected to show that their windows meet the specification. This may mean some modification to their design as against a standard product. Hopefully, there will be a suitable, easily available, window. The successful window manufacturer will provide the construction details in the form of drawings etc. These will become part of the 'as built' contract documentation. This contrasts with the traditional designer's specification which clearly identifies a particular window.

It may be considered that the performance specification relieves the architect of the need to make decisions regarding the details of construction. On the other hand, with the great variety and continuous development of components it is unrealistic to expect an architect necessarily to be aware of all the choices available.

Designing buildings to a performance specification reflects the increase in the complexity of technology. It is inevitable that more people will become involved in the design and construction processes. We are in an ever expanding cyclical movement. As environmental and expectational standards are raised, new technologies are spawned; these in turn lead to new and extending fields of knowledge, which are mastered by others. These then need to be integrated into the design and construction processes. The design process itself expands to accommodate the specialists who are brought in at an early stage. They also have a greater say in how and when their contribution is to be integrated into the system. The requirements of the specialists may determine the construction techniques and methods used. For example, the services and environmental facilities for mainframe computer installations demand specific standards. The enclosure walls and floors have to satisfy these standards. This could mean that the construction is different from that of the rest of the building. What goes into buildings is having a great impact on the building elements.

The traditional procedure for the design process is shown in Fig. 1.7. Compare this with that shown below it for the quasi-traditional procedure. The looping lines show the increase in the interaction between the growing member of people in the construction team. The specialist now has to be in contact with designer and builder.

Another procedural model is becoming common in practice. This takes as its rationale that building technology is the discipline most suited to oversee the construction process from inception to commissioning, to running and maintaining. The ultimate aim is to produce a building that satisfies, absolutely, the client's brief. The part that each member of the building team takes is basically the same; the architect designs; the builder builds etc. but the dialogue of each is changed. In some cases this is extended and with others it is reduced. New actors appear on the stage and some of the bit parts are expanding. The communication network is shown in Fig. 1.8. Nearly all the actors in the team are involved in the developing and staging of the plot, with the cues and prompts perhaps now even more important! Their timing and synchronisation are the responsibility of project management.

The capacity for coping with change is considerable with this network approach to design. As more people are involved at an earlier stage, new ideas can be introduced at the drawing board based on informed data and information. A greater breadth and depth of experience is available to produce a solution which is of optimum value. With this approach a more pragmatic view of technology can be achieved. The focus of attention is on ensuring a high level of buildability, reconcilable with the maintenance of the building's continuing performance.

'Buildability' is simply ensuring that the design of the details and materials and elements does not conflict with the most viable methods of construction. An instance would be the dimensioning

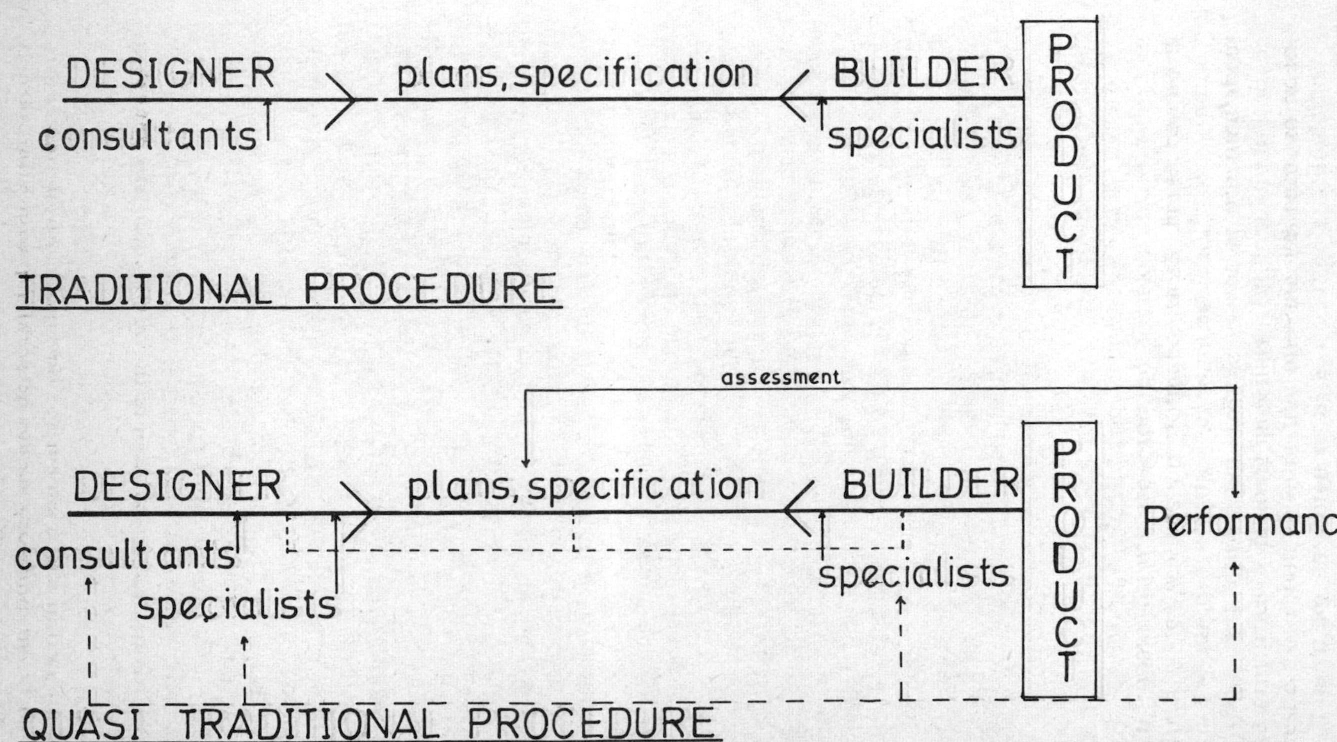

1.7 Precontract procedures

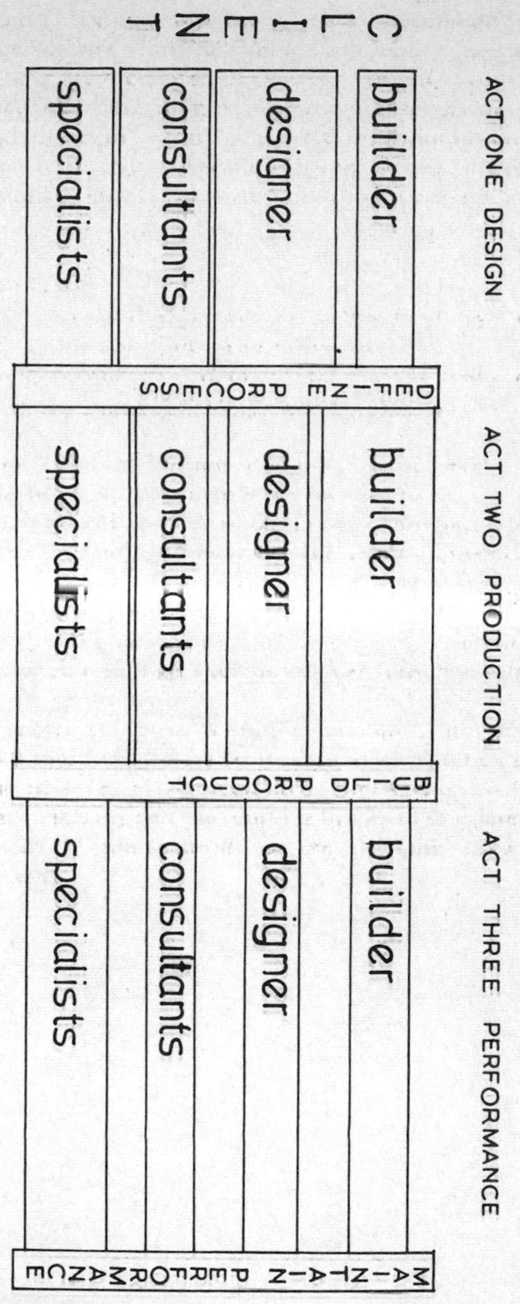

1.8 Network procedure

of brickwork between window openings, say in forming a pier. If a dimension takes into account the length of a brick to reduce the amount of cutting then buildability has been achieved. Creating room sizes which take some account of the size of plasterboards as dry lining will save material and labour time in cutting and fitting. The application of the limitations and capabilities of plant and machinery should be considered at design stage. We shall return to buildability when considering design and production.

7. Technological Change and Development

Living as we do in an ever changing world, each day brings news of new knowledge, scientific breakthroughs and technological developments in all spheres of activities. The rate of change does not appear to have diminished, despite the worldwide recession running in to the 1980s. Not only are changes occurring in the technologies themselves but also in the way that they are managed, as seen in the previous chapter. In one way or another advanced materials or new method (by definition) have an effect on the manner in which they are used.

It is as important to understand the original relationships involved in the adoption of new technologies as to understand their scientific basis. The development is of little use if it cannot be put into practice. Indeed, it can be argued that until it has been put into practice it is not a technology; it remains a theory or idea. Technology has been defined as 'a discourse, treatise, study or practice of the applied (useful) arts and sciences'. Fig. 1.9 provides two illustrations of change relating to building technology. Most technological changes do not develop along one line; usually they are a combination of many disparate developments. Many of the developments in construction have taken place because of research and advances in other industries. The use of plastics, for example, was pioneered elsewhere but adapted for pipework, kitchen finishes and sanitary goods. Electronic control systems have evolved from aircraft, and space craft design for use in environmental control systems.

Some of the factors that have influenced technological change include:

1. research;
2. economic and commercial gain;
3. wider educational opportunities;
4. greater expectations of society.

auger → brace and bit → power drill — able to drill many materials
multi-speed — speeds up hole drilling
percussion — able to take attachments

concrete → reinforcement → diaphragm walls
cement
prestressed
post stressed
precast

bentonite slurry
pumps
clam excavators

deep basements
cut off walls
retaining walls
underpinning
deep foundations

1.9 Illustrations of technological development

1. Research

Research can range from a highly complicated long-term activity based on sophisticated tests etc. to a simple survey to ascertain a point. Some research has closely specified aims, such as producing a material to meet a performance requirement, whereas other kinds are more general in nature in that the outcome is not wholly predictable.

The construction industry draws upon many sources for research and development. These may be established research organisations such as the Building Research Establishment and the Cement and Concrete Association, or individuals working in amenity departments. A building is made from many disparate materials and parts which have evolved from other uses. If the built environment is to keep pace with technological demand it needs to invest heavily in research. Unfortunately, this investment is currently low compared to that in other industries. It has been estimated that 0.5% of the value of annual work in construction is spent on research. This compares with research in other major industries of at least 2%. Appropriate research into industrialised building systems, prior to their mass adoption in the 1960s, might have avoided the estimated repair bill of £2,500 million.

One example of a well-researched and developed building system is CLASP (Consortium of Local Authorities Special Project). Time and energy was spent in the design stage, prototypes were carefully monitored for performance and then further modified in the light of experience. Reports are published by the consortium chronicling progress.

One argument put forward is that it is impractical to test complete systems as this needs to be done over a lengthy period of time. To build an innovative structure may cost a few million pounds. Who pays for this? For how long do we assess it? Who uses it and carries out the assessment? While with consumer goods such as cars, manufacturers expect to make many of the same for a mass market and the research and development costs can be distributed over a large total number of units, making only a small addition to the selling price, this is not the case in building.

An interesting situation has arisen with regard to the use of timber-framed structures in housing. Only a little detailed research has been done relating to their performance. In the early 1980s thousands of timber-frame houses were built without much appreciation of their long and short-term behaviour and the standards required. Has the industry created a time bomb of rotting structural frames? Timber-frame construction was a very suitable case for detailed research and development before its wholesale adoption under UK climatic conditions.

It appears to be a failing of the construction industry that it cannot confidently predict or ensure the behaviour of materials and components in use. The Agrément Board was set up as an assessment body, to give credence to components and materials as regards their everyday and long-term performance. The number of products certificated each year is exceedingly small when compared to the total output. As Atkinson has pointed out, there is no body which monitors technological change and development in the construction industry, let alone gives direction or lists priorities (G. Atkinson, Change for the Best? *Building*, 13 October 1978).

2. Economic and commercial gain

Much change and development is powered by the need to produce something better than before, and in the process to capture a market and thereby make economic gain. Most research producing new ideas is financed by commerce and industry. It may be directly funded, through the firm's research department, or indirectly; by contribution to an independent research association or by funding specific projects.

In the main it is the manufacturers of components and materials who lead development and who seek markets for their products. Builders operate in a small, one-off type of market and do not expect to be able to originate new products themselves. Where builders have invested in new systems the return on capital and project has not been high. Even in the boom time of the high rise industrialised systems there were few builders who managed to make reasonable profits on their innovations. Indeed, many went into liquidation within a short time.

A large market is required in order to sustain heavy capital investment. It was not until 1970 that industrialised housing units became cheaper than traditional built. In 1967, at the height of the use of industrialised systems, the comparative cost comparisons per square foot were:

traditional	£3.95
industrialised	£3.80.

In 1972, when industrialised systems had all but ceased, the comparison was:

traditional	£5.08
industrialised	£5.45.

In neither year was there a commanding advantage for either method (*Housing and Construction Statistics*, HMSO, 1973).

The major advances have been made by manufacturers rather than builders, as Bowley has shown (*The British Construction Industry*, 1966 – although now somewhat dated his findings are still pertinent).

3. Wider educational opportunities

As the population becomes better educated it generates its own expansion. The general level of competence rises and what were advanced ideas soon become the norm. This creates room for further advances. A well educated and trained working population is at the root of a country's prosperity. The report *Competence and Competition* (NEDO, 1984) compares provision for industrial training and education with that of the UK's economic competitors. It shows that spending on education and training in the UK is half that of West Germany. This is correlated to the strength of that economy as against that of the UK. Nations with buoyant economies believe that their levels of education and training in industry are a major contribution. A further example of the need to provide trained people is seen in the number of engineers that are produced in Japan; there are ten times as many as in the UK, with a population only twice as great. There is good evidence of their economic activity worldwide. A survey carried out by *The Sunday Times* (25 November 1984) asked industrialists about the level of skills they wanted from employees for all functions in the company. They said that even in a time of high unemployment there was a definite shortage of skilled personnel. They put this down to a lack of provision for training at post-school stage. This was reinforced by a further survey carried out by *The Sunday Times* in April 1986.

Although in the UK there has been an expansion in education at higher levels following the Robbins report of 1963, it is argued that this needs to be more widespread. Now there is the realisation that education and training cannot stop after qualification. Training needs to be continuous through a working life. Indeed, retraining may be necessary in order to cope with new jobs or roles. This matter is now treated seriously by the professional institutes within the construction industry, under the concept of 'continuing professional development' (CPD).

Nevertheless, technological developments have taken place. Industry itself needs to play a major role in providing the opportunity for structured in-house or external updating courses for all its employees. In the case of the professional institutions they expect their corporate members to undertake some specific activities that reflect their continuing interest and updating in matters relating to construction. Some institutes set a minimum number of hours for such activities each year and require a record to be kept.

More and more clients ask for details of the qualifications and experience of the management teams for their projects. They see it as being in their interest to have the highest levels of training and up-to-date experience for the efficient and timely completion of building work.

As D.S. Landes (*The Unbound Prometheus*, 1969) has shown, in spite of the clear corollary between education and industrial progress, there is still a strong residue of resentment and apathy amongst industrialists towards technical training and education. Here Landes recalls late nineteenth century attitudes which unfortunately are perpetuated today, perhaps not consciously but in the lack of good practice:

> As we have seen, even elementary education encountered suspicion and resistance in England; *a fortiori* technical instruction. There were those industrialists who feared it would lead to a disclosure of or diminish the value of trade secrets. Many felt that 'book learning' was not only misleading but had the disadvantage of instilling in its beneficiaries or victims – depending on point of view – an exaggerated sense of their own merit and intelligence. Here management was joined by foremen and master craftsmen, who, products of on-the-job apprenticeship, despised or feared – in any case resented – the skills and knowledge of the school trained technician. Still other employers could not see spending money on anything that did not yield an immediate return, the more so as the notions imparted by these classes and institutes invariably called for new outlays of capital.

Such attitudes exist in the construction industry today, as is evinced by the small numbers of graduates that enter general building each year. In 1983 it was approximately 400. A restricted survey of building firms' attitudes to the employment of building graduates found that there was a reluctance to accept that a degree student had anything to offer, that 'book learning' was not valued (I.E. Chandler, MA Thesis, University of Sussex 1979).

4. Greater expectations of society

Society is demanding increased comfort and a wider variety of accommodation. Schools, hospitals, factories, houses are being built to ever higher standards. We constantly seek to improve our surroundings, whether internal or external. In order to accommodate sophisticated plant and machinery, buildings themselves take on high technology construction and services. Building technology has to be developed to meet the demands of advanced technologically based industrial processes. Internal environments need to be carefully created to ensure the ideal conditions for manufacturing processes, whether these be the making of computer chips or the brewing of beer.

Most new houses have central heating installed as a standard utility. Future demands may require a heating system that does not depend upon energy sources as gas and electricity, e.g. heat pumps.

The general expectations of people are also being raised as regards wealth and consumer products. Demand is not based solely on basic human needs. It has been recognised that the definition of poverty has to be different from that laid down in the 1950s and 60s. In the UK you are now 'poor' if your income is below a certain level and you do not possess a washing machine or television set.

It can be seen that technological change and development are not the result of one set of factors; they are grounded on man's desire to improve wherever possible. With an increase in surplus wealth over and above the basic requirements for existence, the possibilities of development are, in theory, infinite. Whether or not society can cope with them is another matter. This is an issue that runs through the next contextual factor, the society/technology interface.

8. The Society/Technology Interface

This contextual factor can be seen as an all-embracing essence that permeates and influences every action in the creation and maintenance of the built environment (see Fig. 1.4., p. 15). Without our present state of society there would be no technology; if there was not this level of technology society would be very different.

As we develop and use technology there are moral and ethical issues which are brought into focus. For example in the early 1980s there were many office develpments in and around London. New blocks came onto the market daily, yet there was obviously an overprovision. Space was unlet in existing properties and new blocks could not find tenants. It was estimated that there was upwards of 2,000,000 square feet unlet. Some space was unlet for years, not because it was unattractive or expensive, but because the economy was not providing the expansion for the creation of office jobs. With the advent of information technology it has been said that the numbers of people employed in offices will be reduced as computers will carry out functions more efficiently. If this is the case then why are these buildings being erected? In the many answers to this question lie the issues that modern society is now debating. Before the 1960s there was little public debate on such issues: nowadays the public concern is reflected (in one measure) by the number of public inquiries relating to the proposed construction of buildings. Some argue that this is a waste of time and money and that the objectors are causing trouble for the sake of it. A public opinion survey carried out by Marplan in September 1984 found that over 80% of the respondents would break the law in protesting against what they thought an infringement of their rights or property. If this is a fair reflection of people's attitudes, then the level of awareness and concern is very high. Public opinion is capable of changing and modifying proposals in any field of activity.

There was much comment about the appearance of concrete as a material for the facades of buildings. It was criticised as being grim and unattractive; it tended to stain with age, and proved an ideal blackboard for graffiti. Consequently, the use of concrete for this application has virtually ceased.

Concern with health risks has led to the replacement of material and components based on asbestos. No new components containing asbestos are now fixed in buildings. Alternative materials and construction details are now required. It was medical research, trade union concern for members and public opinion which saw the demise of this material. Its reputation is such that buildings are evacuated during its replacement.

Another situation where note has to be taken of public opinion is during the construction process. Where buildings are constructed in crowded town centres they can cause a nuisance to adjoining users. Machinery noise levels might be unacceptable, even though complying with present legislation, in which case alternatives might have to be used or work patterns modified.

Local residents may object to the constant flow of vehicles to and from the site, which again might cause a rethink on the construction methods adopted.

The general views and attitudes of society towards technology, and building technology in particular, are subject to change. For example, the public at first acquiesced in the adoption of industrialised building systems. Handovers and openings were attended by high ranking government ministers and members of royalty. Such events made television and newspaper headlines. The public acclaimed the new technologies. Now it is a different story, with instead a flood of disclaimers of responsibility for the failure of this technology.

As previously mentioned the interface between society and technology can be made and broken at will. Only if that technology is seen to improve some aspect of life will it be accepted: society is us and we are our society. In the UK we have grown up accustomed to technology – technological subjects are part of the school curriculum, and are taught in technical colleges, universities and polytechnics – but we cannot yet measure or foresee what influence information technology will have on society, especially in the construction industry. Will computers be able to select an optimum solution? The day will soon come when every site will have a desktop computer with all information relating to a project literally at one's fingertips. How will this affect the technology?

9. User Values

Most of what we value – such as love, sentiment, enjoyment, self-satisfaction, achievement, status – cannot be measured in absolute terms, monetary or otherwise. What may have great worth for one person may have none at all for another. A more reliable measure of value than money is utility. For instance, a spark plug is of little use until it is placed in an engine block; some small part may have little intrinsic value, but its failure could create a disaster.

The parties within construction have sets of values which can differ widely from each other. Fig. 1.10 gives some values which may be associated with the four main groupings.

Contributors to the volume edited by Hutton and Devonald (*Value in Building*, 1973) have identified five approaches in the quest for a satisfactory measurement of value in the built environment:

Anstey:	value as the power to serve man;
Stone:	value as a relationship of benefits;
Fleming:	value as a basis for choice;
Markus:	value as an aspect of performance and evaluation;
Bunstone:	value as a measure of efficiency

These approaches will be explored further when we look at some of the issues in Part Two of this book, on Design.

Building is primarily a service industry, despite its manufacturing and production overtones. Buildings are produced primarily to serve human beings in the pursuit of their activities. Ultimately, it is the value which users give to a building which will determine its worth. A structure which may have value to the construction team is no good unless it is effectively used when built. Therefore, it is important to be aware of the different trends, fashions and values of society. In this contextual framework factor we take a different perspective to that adopted in discussing the first factor, Function.

Fig. 1.11 shows how user values are translated into building types. As society is in a dynamic state it is inevitable that the values will change. In addition different socio-economic groups will display varying values. Worth is now given to features, equipment and the environment which were not considered a few years ago. A prime example is double glazing. To install the average system on the basis of a monetary saving due to reduced fuel expenditure is to delay benefit for up to 20 years. It is only after this period of time that the cost of installation is covered by the savings from the reduction of heat loss through the glass. Many people go to the trouble of having old windows replaced or modified to double glazing for the following reasons:

(a) as old or rotting window frames need replacement one might as well fix double glazing;
(b) to cut out draughts;
(c) to give some noise reduction from external sources;
(d) to make the house look 'up to date';
(e) to enhance the 'saleability' of the property;

CLIENT	investment aid to production status/prestige durability marketability running costs
DESIGNER	creativity reputation performance buildability
BUILDER	quality attainment profit continuity of work repeat work
PUBLIC	meets needs non-pollutant aesthetically pleasing non-disruptive

1.10 Values: contract

None of the above give a direct financial benefit but they do give user satisfaction. Equally, an increase might occur in the installation of heat pumps as a more efficient method of heating, thereby creating a better cost benefit to the user, with a subsequent fall in replacement double glazing.

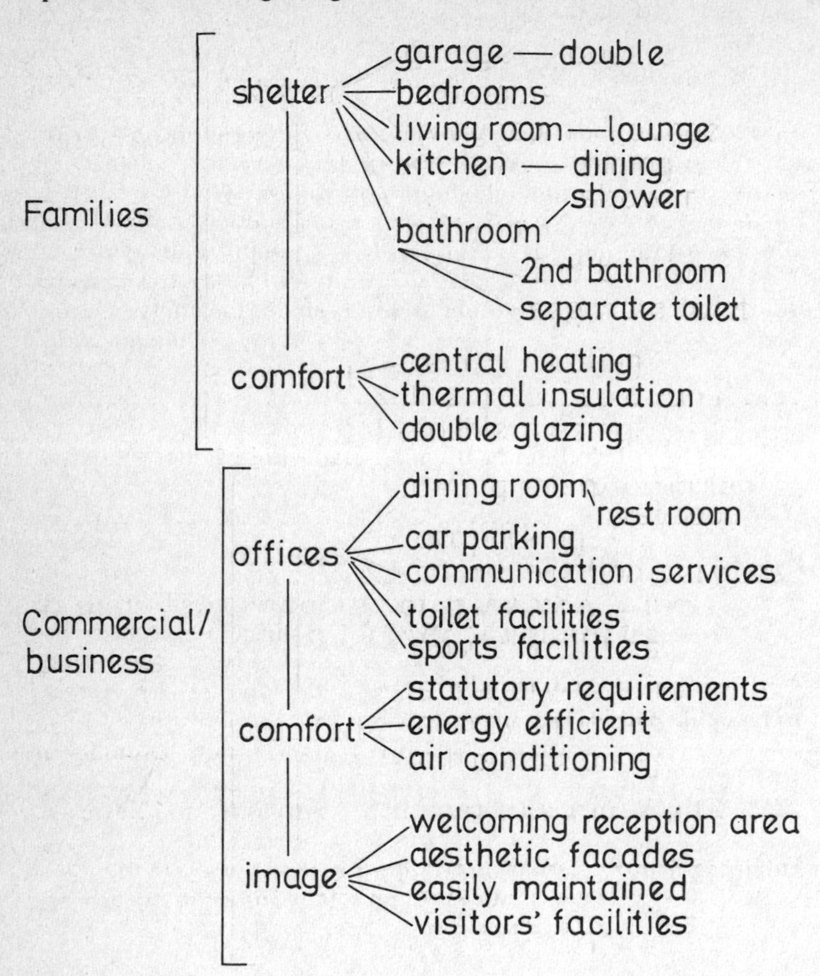

1.11 User values, some examples

SUMMARY

In this Part some of the everyday environmental factors which have been identified have, to a lesser or greater degree, moulded the technological solutions used at all stages in the construction of buildings. No technical solution has been plucked from the air, it has been determined, consciously or unconsciously, within the existing ambience of its culture. Here we use the word 'culture' in its widest sense. The UK has a culture which is different to that of the USA, or Africa or Asia. Nevertheless all could be using the same industrial resources but deriving very different solutions. It will be noticed that many of the questions posed remain unanswered. This is intentional! As a function of the nature of the problems, what might be a right answer here and now may not be appropriate tomorrow. What is needed is the ability to ask the right questions. Hopefully, this Part has created awareness and provided some pointers to where the answers might be sought. Bringing these elements together can produce the technological solution which is optimal for the circumstances.

QUESTIONS

1. Taking a common construction detail show how the conceptual framework factors have influenced its choice.
2. Discuss how public opinion affected the technology of private housing with regard to timber-frame construction in the mid-1980s.
3. Should there be more legislation affecting the technology of buildings, or are the present regulations sufficient and pertinent?
4. Illustrate an example of a technological development from another industry which has directly influenced a construction detail.
5. Discuss whether technology is affected by the need to produce viable solutions or is motivated primarily by monetary gain.
6. Discuss the extent to which the structure of the industry will influence the technology used.
7. To what extent should the builder be aware of clients' values and expectations?
8. Discuss the implications for building technology if there were to be a far more rigid control of safety aspects, both during and after construction.

Part Two
DESIGN

1. Introduction

In this Part the various factors relating to the process and issues, connected with the design of buildings will be explored. We shall seek to discover how design is an integral part of the construction of any building. The techniques and considerations taken into account by designers will be described and related to the technology. The chapter on Value will clarify the viewpoints of the parties to the construction process. Techniques such as value analysis will be described and related to human factors, performance specifications, life cycle costing, energy accounting etc.

The relationship between design and construction will be considered first and then followed up in the next Part on Produotion.

A building failure is described to illustrate some of the problems in design and a brief description of the use of computers in design is given, together with some thoughts for the future. Finally, the summary will highlight the factors discussed and investigate some of the constraints affecting design.

In the design process there must be an objective. An ultimate goal must be sought, which may or may not be reached. In trying to reach this objective a route can be planned and its stages identified.

Alternatively, random thoughts and activities may lead to a solution. Just such a process is exemplified by Clive Sinclair's solution to the problem of producing a relatively flat and small television screen. To create the picture a beam has to be projected over a minimum distance; this makes the depth of the tube quite long. So how to reduce the distance? It came to Sinclair whilst bathing: to turn the beam through a right angle so that it would travel the same distance but within a reduced depth of the television set. This enabled the objective, a very small TV set, to be made. The design problem was known as well as the objective but in this case a systematic approach to reaching the solution was inappropriate. This approach could also be used in building technology, but it is usually found within an outline of a method. A strategy is set out. Intuition and flashes of inspiration play their part but these cannot be relied upon for 'every-day' solutions. Generally, the process of design interweaves through six identifiable modes. These are shown in Fig. 2.1. It is not a linear process;

much discarding, rejection and readoption, and interweaving and backtracking will be necessary as the design takes shape. Any design – to be worthwhile – should start from first principles. This is not to say that the wheel is to be reinvented everytime but rather that the basic principles ought to be considered.

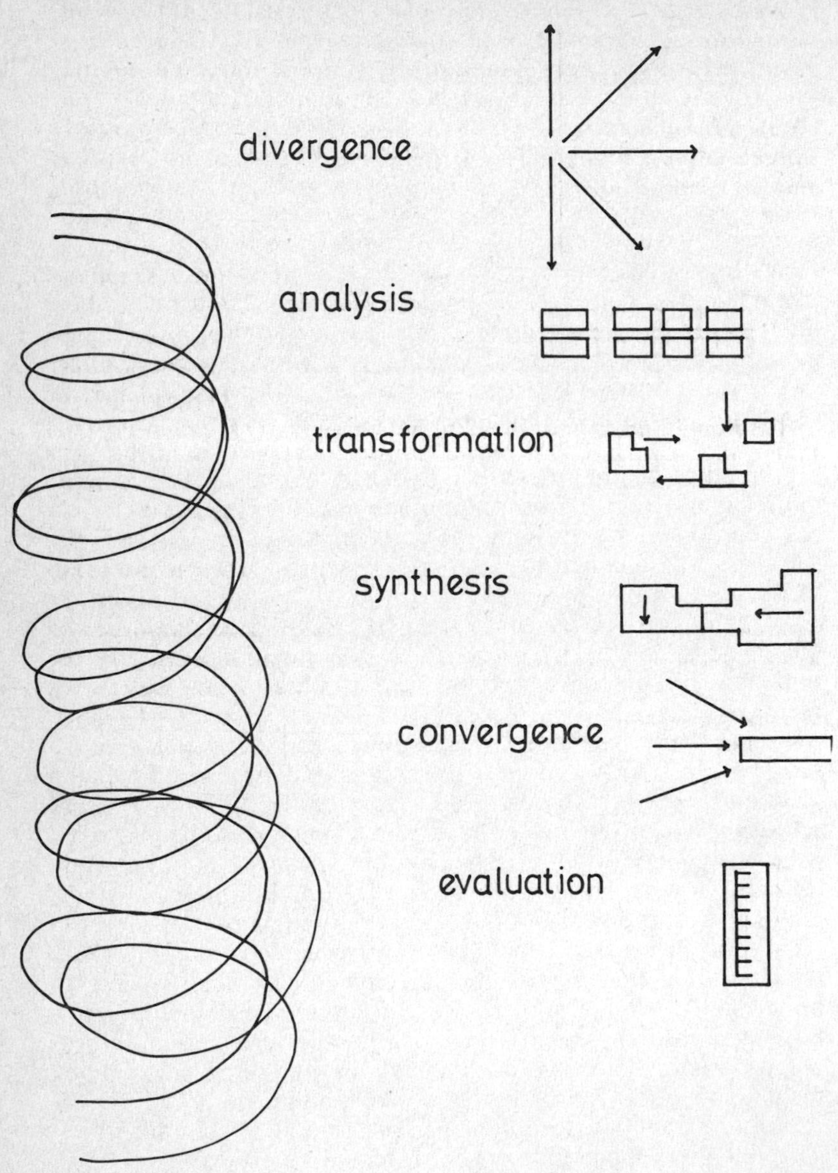

2.1 Design modes

Even the design of a traditional house should reflect this process, although the ultimate solution may resemble many others already built. Unfortunately, experience and research by the BRE (Digest 268, *Common Defects in Low Rise Housing*) has shown that many buildings suffer from common errors which are attributable to design. This indicates that the evaluation stage of design is perhaps not very effective, perhaps due to a natural reluctance publicly to admit mistakes. Returning to the design modes shown in Fig. 2.1 it can be seen that in the first instance many ideas and solutions may present themselves, which are then analysed in turn to ascertain their exact nature. During the transformation mode a start is made in considering the conceptual framework factors as they relate to each possibility: we have to set the solutions in the real world. In synthesis we are moving towards a reducing number of possible solutions and can now begin to quantify and qualify the options. Greater attention is given to the constraints and they are fully incorporated into the design. Hopefully, one then arrives at a convergence which points to the optimum solution. If possible, an initial evaluation should then be undertaken before a final commitment is made. This can take the form of computer-based models, reference to histories of similar structures or a complete reappraisal of the whole project; a prototype might even be constructed. In practice such evaluation often takes place during construction, and although this late is better than never, it will interfere with the construction process. If changes and modifications are made then which are not due to unavoidable circumstances they must cast some doubt on the thoroughness of the initial design process.

The design process must have objectives; the main one is the provision of a building or structure. But there can be other objectives which impinge upon and perhaps direct the final outcome. If the building is required for prestigious purposes then the architectural style and appearance may be the dominating element of the design. An advanced factory will need to have a greater degree of flexibility in its ability to accommodate unknown manufacturing processes. High energy costs could focus the design criteria upon the need to produce a passive heating system within the external walls.

A major objective is to build to the lowest cost whilst still meeting the specification. In many cases only a certain amount of money is available and the design must be tailored to fit the budget. In the first instance the designer will need to meet the demands of the client's brief. This brief ought to be very clear with respect to the client's expectations, but the designer will be expected to advise the client where the demands might be unrealistic. In any event most solutions are usually a compromise. Here it is opportune to mention the involvement of other parties to the design process.

With the increasing complexity of technology, the speed of change of that technology and the subsequent range of specialist discipline, it is essential to involve these interests at an early stage. Without such participation the design process would stretch through time with constant revisions to cater for the next-in-line specialist's contribution (the builder being one of these specialists). The work on site can be disrupted if all the ramifications of the disparate technologies have not been fully integrated into the design.

There is a paradox here which needs careful understanding and monitoring. On the one hand there is a trend towards narrower and narrower fields of expertise, each with its own practitioners. On the other hand the need for the total integration of these into the design and construction of a building demands simultaneously a very wide view. Instead of the specialist making the design process easier he can make it much more difficult in an organisational sense. This strikes at the core of any technology, especially building technology. A common definition of technology is 'the practice of the applied (useful) arts and sciences'. The keyword here is 'practice'. It is marriage between the arts and sciences in order to produce a functional and efficient building. The process of this marriage is constrained by the structure of the industry; some of these issues and their influence on technology will be considered towards the end of this Part. In design the aim is to take the best of the arts and to produce a solution which is rooted in scientific knowledge.

2. Value

Value is nearly always relative and in the final analysis is a matter for personal or group judgement. It is difficult to quantify, although many people express it in monetary terms. Another way is to consider the benefits or otherwise of a course of action. For example, the value of a hospital can be placed over and above the cost of its construction, running maintenance and staffing. At the other extreme the cost of a folly could far exceed its value as decoration. What most people want is simply 'value of money'. If in the production of a building or the provision of a service the reward matches the input, then most people will be satisfied. Where a mismatch occurs one party may benefit at the expense of the other. If a client discovers he has been overcharged, he will be resentful, while the recipient of the overcharging will walk joyfully to the bank. Should this be blatant or occur often distrust will ensue, with a consequent deterioration in relationships. The reverse situation, for example where a worker is not getting paid at the level expected for the job, will also lead to a decline in the standards of the relationships, with a possible decline in the standards of the job. It may be that, historically, a mutual distrust has grown between client and builder which has led to the growth of the independent professional acting on behalf of the client. This past distrust is now perpetuated in the clearly defined roles and functions of architect as manager and quantity surveyor as cost controller. The cost control of a contract is monitored by two parties, each looking over the shoulder of the other. A study for the RICS by Reading University on the relative costs of building in the UK and the USA found that in the UK costs were 10% to 15% higher, a difference which could be attributed to the fees of the quantity surveyor.

Returning to value it has been seen in Fig. 1.10 (p. 53) that each party to the construction process brings a set of different values to the production of a building. What they must all agree upon is a common definition of value pertaining to that particular building at that particular time. Another building in another place and time could have a very different value. Each project is unique and the value criteria will need to be reassessed for each and be made known to all those concerned.

CLIENTS' VALUES

The client's interest in value depends to some extent on the type of building, what it is for, and the purpose of its construction. If the building is a commercial development, i.e. for renting, then the value to the client/owner is primarily one of investment and income. It has to be designed to ensure that it is lettable and that the floor areas are maximised, as rents are based on the gross floor area available to the tenant. The lease issued to prospective tenants may have clauses which state that they are responsible for the maintenance of the external fabric, together with internal fixtures and fittings. They may also be expected to contribute to common services such as lifts and heating. In this case the client/owner is not over concerned about the durability of the external fabric, as this is the responsibility of the tenant. The main value of the building to this type of client lies in its attraction to prospective tenants. Initial costs will be kept to a minimum and the emphasis will be on ensuring that the building lasts long enough to reap a return on capital and make a profit. There is no doubt that it will be in the client's interest to reduce the possibility of excessive maintenance to the building in order that tenants are not put off taking up leases. The client must meet the demands of the market. A recent instance illustrates how costs, in the eyes of the client, may be regarded as excessive. A large building was to be refurbished in the City of London. The client considered air conditioning to be unnecessary in relation to the number of times it would be used in the average London summer. Also, there would be problem in fitting it into an old building fabric. The client's property market advisers said that they would have great difficulty in letting the building without air conditioning; the market demanded this facility in that situation. This resulted in a 10% increase in the construction costs of the building, with not a particularly significant increase in the rents charged.

If the extreme situation is taken where a relatively short life material or component is used in order to reduce capital cost, then the implication for technology is profound. One needs to be confident of the life of that component, however long or short. Interim maintenance requirements need to be ascertained and made known to the tenant. The designer will be restricted in choice, as there is an upper limit on cost. With short life materials and components the chance of premature failure can be high. Extra care and attention are required during the construction stage to ensure proper fit in order to reduce the possibility of failure.

Another possible consequence arising from the client building for investment is that only tried and tested technologies will be used. It is a natural desire to play safe in appearance, services and accommodation. The use of new technologies may create apprehension in prospective tenants faced with unfamiliar styles and the possible risk of technical failure.

If the buildings are to be used as an aid to manufacturing production then they can become a secondary interest where the client is the owner. The need is for a building that will satisfy a set of criteria determined by the way that the building is to be used. Although factories tend to be viewed as being for one all-embracing use the variety of manufacturing/assembling processes carried out in each is legion. Each factory should be designed to accommodate fully those particular processes. Where advanced factories are constructed these need to be extremely flexible in their provisions and layout. In the case where clients are looking to build for themselves there will be a greater scope for control over the exact requirements. Indeed, many industrial processes require special structures and facilities that can only be housed in specially constructed units. The technology for these tends to be as advanced as the plant and equipment they contain. Many industrial processes require stable environments, dictating that the structure, fabric and services be of a high specification. With this type of 'one-off' building there is a strong possibility that its life is closely associated with the life of the process it accommodates. When that process becomes redundant so too does the building. In that case it is to be hoped its life will not be shorter than the time required to recoup the costs.

As this type of building is used it must contribute to the efficiency of the internal processes which in turn earn a surplus to replace the building in the future. General technological change is occurring at a rapid rate and a dilemma arises between building for a long life and meeting future needs so far not known. It is quite possible for some industrial processes to be outdated within ten to fifteen years of their inception. This is a relatively short life span for a building. The client needs to be well aware of these factors and to incorporate the information in the brief for the designer. The design will have to deal with the demands of present requirements with a view to either planned obsolescence or major readaptations.

To create for customers and the public alike an appearance of stability, wealth, confidence and integrity a client may want a prestigious building. An eminent architect may be employed to give further credibility. This architect might be of a modern school who is prepared to explore and put into practice new ideas and technologies. There are certain fashions in styles and facades, for example the common use of bronze-tinted glass in the 1970s and the use of atria in the 1980s. The tinted glass did provide a functional requirement, that of reducing solar gain and glare and giving some privacy to the users.

There is a degree of 'one upmanship' involved in commissioning

The perspectives that architects may bring to design have been introduced earlier in Chapter 6 of Part One, dealing with the approach to design. Here, one or a combination of ways of seeing architecture were said to be the basis of organising and structuring design tasks. These provide a means by which the design in its finished state can be viewed. The intention of the architect could be identified in any building, given some clues. Within this language of ways of seeing buildings is a grammar; the rules, stages, etc. which produce the language. The varying emphasis that each designer puts on these aspects will reflect his values. The product will be a realisation of the way of seeing the design process. Fig. 2.2 shows six different design processes. Numbers 1, 2 and 6 are primarily cognitive approaches, whilst 3, 4 and 5 are orientated to the manipulative. One may compare numbers 2 and 5: in 2 the words 'identify', 'analyse' and 'communicate' are used; these refer in general terms to processes of the mind. The emphasis is on how the designer organises his own conceptual approach to the design. On the other hand in 5 the emphasis is on the stages that can be identified externally and practically. There is also greater weight placed upon activities in the later stages of the design process. In the case of the RIBA design process (the plan of work) this emphasis may give a clue to the reasons for the common practice of revising details, materials etc. during the course of the construction process. If the architect is encouraged to spend time, energy and thought on these latter stages it may be that the former suffer. In other words not enough attention is given to the first three, so that a poorly thought-through scheme might result. There might be an assumption that the design is only finalised during the latter phases of the process and that it is here that alterations can be made. Is it significant that compared to the five-step design process there is no 'evaluation' stage?

It is also important to distinguish whether a design process is inductive or deductive. With an inductive approach the process starts with the details: each is considered and designed and then added to the others to build up a complete solution. With a deductive approach the overall intention or idea is first realised and the details grow from this.

For instance, if the inductive approach is predominant, then details such as window openings, doors, structural openings, joints etc. will be the basis of the designer's concern.

With a deductive approach, there will be greater concern with the overall effect and ambience of the building. So long as the desired intentions are realised the choice of detail becomes secondary.

process	design process	supplementary services	process	services	to Architectural Design
initiation imbalance	identify the problem		problem identification	Inception	basic definition preliminary programme
preparation	collect information analyse information	predesign services	analysis of user needs programming	Feasibility	investigation analysis programme abstraction
proposal making	creative leap work out solution	schematic design design development	design synthesis	Outline proposals Schematic design Detail design	synthesis and development volumetric design proposal
evaluation	test solution		selecting from alternatives		reevaluation and modification
action	communicate and implement	contract documents building administration of contract post-design services	implementation	production information bills of quantity tender action project planning operation on site completion feedback	
1	2	3	4	5	6

2.2 Selection of views of design process

BUILDER VALUES

The highest value should be placed upon the achievement of the standards prescribed by the client or architect. In the final analysis it is the builder who is providing a service to the client. Only where a building firm retains the structure for its own use can the aspect of service be reduced. Most buildings are constructed for others, so although a product is the final outcome, it is the service element within the process upon which the builder is most often rated. The builder will be asked such questions as: has the quality standard been met? was the work completed on time? was it carried out in a satisfactory manner? was it done to the price? If the answers to these questions are positive, then it is likely that repeat work will be forthcoming. Building contractors with either a high local or specialist reputation gain upwards of 60% of their work from previous customers.

Perhaps the most important factor for the builder is to ensure a continuity of work. The construction industry has always been susceptible to the vagaries of the economy and government policy, and to a limited extent to shifts in the type of work. On a national macro-economic scale the industry is used as an economic regulator. When a government is endeavouring to expand the economy a simple method is to provide more money for capital expenditure on buildings, whether houses, schools or factories. During times of recession or deliberate contraction there will be less public money available, as in the early 1980s. Together with the decrease in public capital expenditure there is often a concurrent reluctance by business concerns and others to invest, or reinvest in buildings. A recession brings about a loss of confidence in the future which undermines those projects which may be riskier than average. As this pattern of stop-go has been repeated many times since the 1940s uncertainty has become a fact of life for the builder. The Tavistock report, *Interdependence and Uncertainty* (1963), highlighted this problem as it affects the construction industry.

As there is so much uncertainty great store is set on work being continuous. The ideal for most builders would be the ability to move the workforce from job to job as and when demanded without wasting time in between. As it is, layoffs are frequent, even in relatively stable economic times, and operatives have to be rehired for new jobs. This has led to a massive increase in the number of labour-only subcontractors and the self-employed. In 1984 these increased by 13%.

Where a builder has a directly employed workforce employees must be kept in continuous productive work. If they are not actually working then they are entitled to a basic wage. If there is little chance of further work then redundancy proceedings have to be implemented, causing further monetary loss to employee and employer. It can be seen that this is reason enough for the builder to be reluctant to keep direct labour.

On the other hand the use of sub-contractors does have implications for the approach to the technology of building. In the first place there is a threat to standards of workmanship and service during and immediately after the construction process. Control of self-employed labour is more difficult than of direct labour, especially if the builder is responsible for the materials. The degree of supervision over sub-contractors also has to be more intense and far-ranging.

In relation to the above points it is appropriate to consider the value that builders give to completing on time. A major report by the Building Economic Development Council, *Faster Building for Industry* (NEDO, 1983) found that the use of sub-contractors did not by itself speed up the work processes.

> The use of sub-contractors was much more extensive in the south of England than in the rest of the country. Of the projects studied in the south, over three quarters were wholly or mainly sub-contracted compared to only about a third in the rest of the country. Construction times outside southern England were generally faster. As this shows, projects that were carried out mainly by direct labour were, on the whole, faster than using sub-contractors.

The report goes on to say that the use of sub-contractors

> had the most far-reaching affect on construction times overall: only 1 in 10 of the projects on which all or most trades were subcontracted had faster construction times compared to 1 in 3 of those which used directly employed labour for most of the work.

In Chapter 4 of this Part, on Production, this aspect will be investigated further. It is chastening to realise that while it is well within the builder's scale of values to give a service and complete on time, in a great number of instances this is not achieved.

Last, but not least, there is the need to produce a profit, over and above any surplus for reinvestment, expansion etc. Those building organisations which are not expected to make a surplus, such as hospital direct labour departments, need to give value for money. They are also judged against their private sector counterparts as to estimates and final costs. This direct comparison is now required of local authority direct labour organisations.

The value given to making a profit is very high within commercial building organisations. There is constant reference to this during the process and on final completion of the job. Generally cost control is highly structured and a balance sheet can be produced on a monthly basis for each job. Depending upon the level of pricing for the job, will be decisions relating to the type and scope of the technology. On a job obtained by a low competitive tender there will be a risk of cutting corners to make money to increase the profit margin. A job priced on a fair value for money basis will not necessarily suffer from cost cutting, owing to the larger margins for profit. It is not unknown for a builder to skimp concrete thickness in order to save or make money!

Most builders succeed in maintaining moral and ethical standards and still make a profit. If a builder cannot make the intended profit it is a reflection upon the deficiencies of his organisation. There are, of course, external influences which may disrupt the process over which the builder has no control, and this may lead to a loss on the job. However, it is generally within the capability of a builder to turn a loss into a profit or to at least achieve a breakeven point, without a detrimental effect on quality. A case study in *Faster Building for Industry* describes how a contractor re-programmed the original 65 week duration halfway through a contract to 52 weeks in order not to make a loss. This required the injection of more site supervision and back-up effort. It was more worthwhile to speed up the process, than to tinker with quality. As there was more site supervision it was probable that quality improved anyway, over and above that already being achieved in the early stages of the job. The main conclusion of the report was that quality did not suffer because of speed.

PUBLIC VALUES

The increasing number of public inquiries relating to building development indicate an upsurge in interest from the general public. At a local level it may be from residents objecting to the construction of a medium-rise unit of flats because it may spoil their view or increase the flow of traffic. At national level there is the example of the Coin Street development in London: people here are objecting to the nature of the development, saying that housing for locals should take precedence over office development, in order to meet social needs. They argue that the area will be deserted at night, leading to the possibility of higher crime rates. The developers' view is that this is a prime building site for the needs of business and that therefore its full commercial value should be realised.

Another example of public opinion causing a rethink on a development is the extension to the National Gallery in London's Trafalgar Square. The concern here was with the appearance of the building, not its use.

In both cases the outcomes will make a difference to the technological solutions adopted. In the Coin Street development, if the objectors proposals were to prevail, then the technology employed would be basically traditional houses and flats in brickwork. These would require relatively unsophisticated services and amenities. Traditional craft skills would predominate. If offices were built they would incorporate high technology services and amenities. Their basic structure would probably rely on a frame with a high preponderance of prefabricated units for floors and walls. More substantial foundations would be required.

The actual process of construction may cause disturbance to the public; since there is now greater scope for serving abatement notices to prevent nuisance, the builder has to be more careful. Noise, dirt, dust, smoke, etc. will not be tolerated for long, if at all. If these are felt to be excessive the public will protest and, if they are not abated, will seek redress from the law. Even traffic flow to the job may have restrictions imposed on it, being permitted only between certain hours or at a rate of no more than a specified number of vehicles per hour, for example.

As the population generally becomes better educated and more aware of its rights so public opinion will have a greater influence over the built environment. It will concern itself not only with appearance but also with function, need, priorities and care during the construction process.

SUMMARY

In this chapter on value we have given a broad view of the meaning of the concept with respect to different perspectives. The issues mentioned are not the only ones to be considered, nor will they necessarily be seen as important in different circumstances. As change takes place in society so will values, which in turn will influence the prevailing technological criteria.

QUESTIONS

1. Discuss the extent to which the values of the architect should override those of the builder as regards technological choice.
2. Should the public have a greater or lesser say in the development of land?
3. Consider your own organisation; what are the values it espouses and are they being achieved?

3. Science and Humanity: Their Influence on Design

At the root of all design considerations for buildings and the environment is the human being. All the internal and external artefacts are produced to accommodate the human body in a biomedical or psychosocial manner. To take a common example; steps are designed and built within certain dimensions to ensure that the rise and going are suitable for people; they should involve the least physical effort in getting from one level to another, should be safe and should appear to be safe. The level and placing of artificial lighting reflects the activity of the people in that space; biomedical needs are foremost. Design prior to the twentieth century was empirical and instinctive. Little or no scientific analysis or data was applied to structure, environment or appearance. Nevertheless, buildings did match, to a certain extent, the perceived needs of the times, with the technology of the times. For example, the need for the artificial provision of an internal heat source has long been known and partially satisfied. The Romans designed and built central heating systems for their domestic dwellings; underfloor hypercausts.

Nowadays scientific research and data are utilised for designs. One such field of study is that of human factors. Perhaps a more familiar term would be 'ergonomics', but that is limited in meaning and practice. Ergonomics looks at the capabilities and limitations of the human body in relation to actions. Human factors takes a wider view: it places human actions in an environment and then considers their performance in that system. Successful design is the achievement of a compatible human/machine interface. Here 'machine' is used in the widest possible sense to include manufactured and constructed objects; including buildings. In buildings these can be the elements or the components, and range from lifts to windows, carpeting to door handles and signs. The human being is part of the total system and a full understanding of his or her attributes will complement the environmental factors. Fig. 2.3 shows the aspects involved in a human factors system. Under each of the aspects listed on the diagram there is now a growing accumulation of data. From this bank of data one can begin to predict how the human being will react or cope in particular environments and also extract the information that is needed to reproduce an environment.

As developments take place in the field of human factors so comparable advances are being made in other quantifying and measuring techniques associated with the design of buildings.

Prior to and during the detailed design stages the value and worth of a particular construction detail can be determined. From present and historic records the cost of materials and the process of construction can be calculated. A detail can be rated under such headings as speed of construction, ease of maintenance, availability of materials etc. From these data alternatives can be formulated and then compared. The quantities of materials can be ascertained and in the final analysis a comparative study can confidently forecast the value of a construction solution when related to long-term considerations. Each part, section or construction detail of the building can be analysed in this manner.

By using such techniques as discounted cash flow and the present worth of money it can be shown how the building will perform on a cost basis over its planned lifetime. Methods are also available for estimating running and maintenance costs. These can be related to the initial costs to give some idea of the total costs of building.

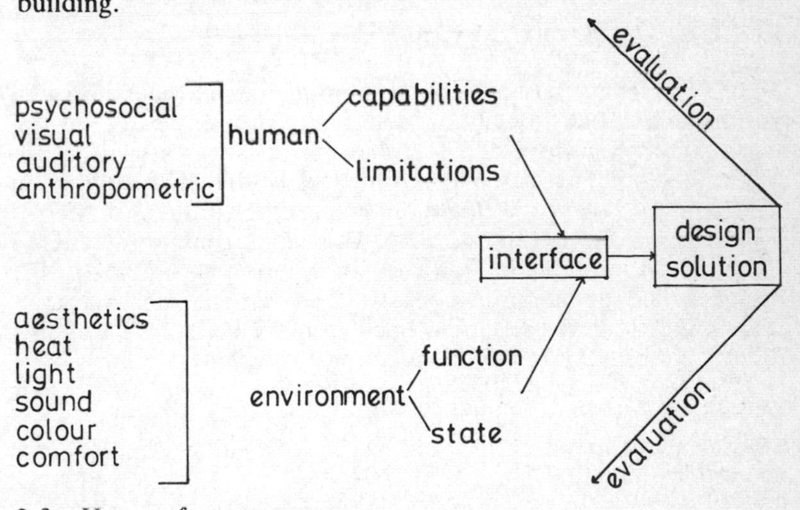

2.3 Human factors: system

Another measurable and quantifiable perspective relating to the building relates to its use of energy. In the total quantity one can consider the energy inputs arising from the manufacture and installation of the building's elements and the parts, together with the energy required to sustain the internal environment. From this can be determined whether or not the building is a high or low energy user.

Buildings can be appraised by their users/owners. Sophisticated questionnaire and data collecting techniques are available to give a picture of a building in use. These can give indications as to its functional efficiency. This appraisal can be linked with the human factors approach.

From the foregoing a number of discrete techniques can be identified. Each relies heavily on scientific data or a body of knowledge upon which decisions can be reached.

1. Human factors analysis, leading ultimately to a *performance specification*.
2. The finding of the worth of a construction detail, carried out using value analysis (*value engineering*).
3. The relationship of initial costs and costs over time, now described by the term *life cycle costing*.
4. The measure of energy, considered within *energy accounting*.
5. To ascertain the performance of buildings the findings of such investigations are related back to human factors (*feedback*).

These will now be considered in greater detail.

1. PERFORMANCE SPECIFICATIONS

Design could be described as an attempt to match expectation with performance. Functionally a designer can be viewed as an information processor (see Fig. 2.4 and 'The use of existing data in evaluating building designs and materials' by R.M. Birse, in *Performance Concept in Building*, Proceedings of the 3rd ASTM/CIB/RILEM Symposium, 1982.) The main content of Birse's paper is an examination of the information available to the designer and an identification of improvements necessary. A major finding of this research, backed up by Freeman ('Building failure patterns and their implications', *Architects Journal* 161 (1975) No. 6 pages 303–308) is that 'in design the major shortcoming is . . . no less than a frequent failure to make use of existing authoritative design guidance'. Nearly all building failures can be attributed to:

(i) failure to collect the right information, in two-thirds of cases;
(ii) failure to act correctly on information, in the remaining one-third.

It can be said that there are ample data sources available but that these are not employed effectively. Referring to Fig. 2.4, it is the failure occurring along the central core that is the cause for

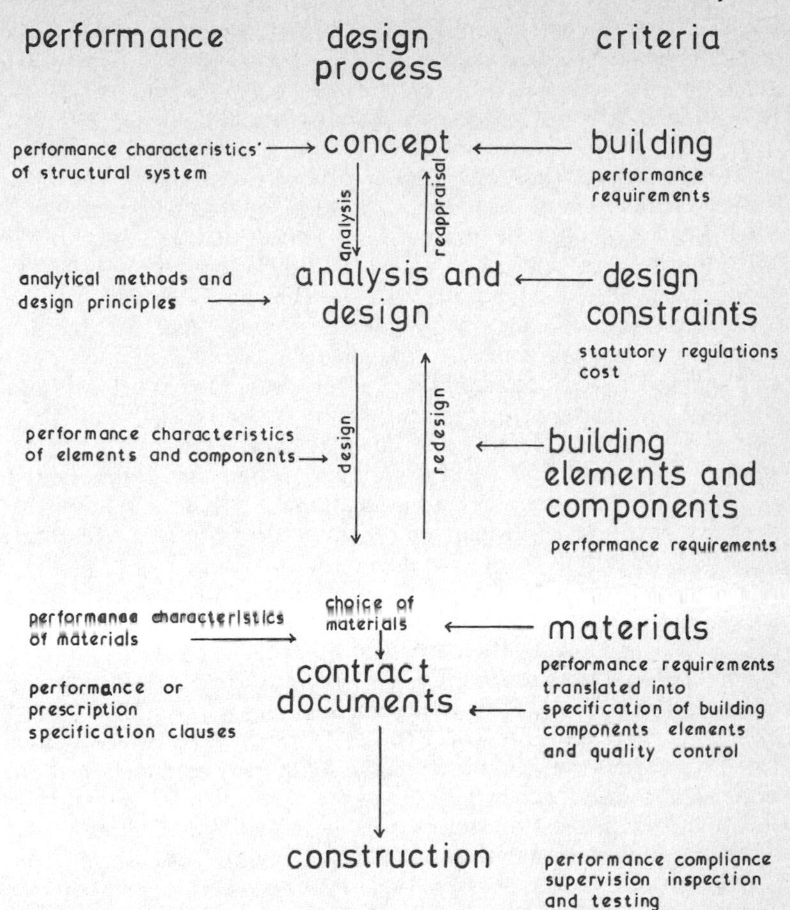

2.4 Relationship of performance and criteria to design process

concern, i.e. the human factor. In using performance criteria the attention of the designer will necessarily be sharply focused on particular requirements, both in detail and in substance. The performance requirements will demand a solution which will provide quantifiable evidence of compliance with them, or otherwise. A mathematical analogy is the simple equation: a set of measurable criteria on one side must be equalled by actual performance on the other.

As a broader analogy of the concept of performance, consider the family choice of motor car. The car must have adequate seating for all the family; the engine must be satisfactory in terms of fuel consumption, speed, acceleration and ease of maintenance. The body must not deteriorate too quickly and the systems such as lights, brakes etc. should be efficient. In the final analysis, from a range of similar models, a choice will be made based only on cost. Aesthetics do not play a major part in the choice, the emphasis being on the functional requirements: they must be met in the first instance.

The common traditional specification for building literally describes the work to be done and the standards to be achieved after the design has been finalised. These standards are often minimum requirements and expressed in such documents as Codes of Practice, British Standards and the Building Regulations. To ensure that the standards are achieved, the designer and builder ought to have copies of the relevant documents, but do they? Perhaps too much reliance is placed on previous experience and memory, with these documents being referred to, but their contents remaining largely unknown (see Mackinder, M., 'The selection and specification of building materials and components: a study in current practice and educational provision.' University of York Institute of Advanced Architectural Studies, Research Paper 17, 1981.)

The clauses in a traditional specification tend to spell out an actual component with respect to what it should look like. For example, a window will be described in terms of its size, material, single/double glazed, type of glass, fixing of glass, finish. In the case of a performance specification the factors would be listed under the following categories (taken from the CIB Master List): general information; composition and manufacture; shape, dimension and weight; general appearance; physical, chemical and biological properties; durability; properties of the working parts, controls etc; working characteristics. Having set down these requirements it is then possible to determine accurately whether or not a particular window can meet them. The performance concept provides a framework within which the desired attributes of a material, component or system can be identified as as to fulfil the requirements of the intended user, without regard to the specific means to be employed in achieving the results. In other words it is a process carried out before any hard design decisions are made: the opposite of a traditional specification. The concept centres on the idea that products, devices, systems or services can be described and their performance measured in terms of user requirements, without regard to their physical characteristics, design or the method of their creation. The key to the development of performance standards is the identification of the significant criteria which characterise the performance expected and the subsequent generation of methodologies for measuring how products, processes or systems meet these criteria. There are various elements within the concept, which are related to the definition of the problem.

(a) Performance requirement

This is a qualitative statement describing a problem for which a solution is sought. It includes identification of: what the nature of the problem is: who has the problem: why the problem exists: when the problem exists.

(b) Performance criteria

Most problems have many facets and require the use of multiple criteria which differ in importance. A performance requirement calls for a solution to be offered. Performance criteria give the set of characteristics that solutions must have.

(c) Evaluative techniques

In order to evaluate alternative solutions it is necessary to have a set of practical measurements that can be used to obtain an 'effectiveness' score for each. A set of tests and measurement techniques are needed which are not biased – they should not favour any one solution. The application of expert judgement may sometimes be the best evaluative technique available. Whatever the case, a public act of measurement is essential to the performance concept.

(d) Performance specification

When a performance requirement has elicited sufficient interest to engender the development of performance criteria and evaluative techniques, then it is likely that an appeal will be made to technology for a solution. This means specifying what is needed in a way that will communicate in the engineering/scientific community. A performance specification comprehends all the information in the underlying requirement and criteria. It also includes the evaluative techniques and identifies the range of scores within which solutions must fall if they are to be considered acceptable.

Performance specifications can be written which are completely unbiased about the means for solving the problem, or can be written so that they are somewhat restrictive about the means without being so narrowly restrictive as a 'hard' specification. (A hard specification is one giving drawings, instructions, materials and the like, i.e. a traditional specification.) An unbiased performance specification can be termed a 'fundamental' specification and one which applied to a family of solutions a 'derived performance' specification.

(e) Performance standards

Performance standards can result from performance specifications. If the measurement techniques are reproducable and the requirements are reasonably common ones a duly constituted body may issue specifications as a standard to be referenced by others, or it may become a *de facto* standard by common usage.

(f) Performance code

A performance code is in effect a performance standard that has been adopted by a regulating body and put into practice in a legal sense.

Performance specifications will now be considered with reference to human factors analysis. Fig. 2.5 shows the linear development of human performance data in relation to performance specifications. The starting point is the human being, with due recognition of his or her capabilities and limitations. These are derived from research and are now well established. The human being can live and act within a range of factors. For example, auditory perception can be measured and limits of hearing, high and low frequencies etc. are known. But in specific circumstances the tolerance ranges may vary. For example, in a factory environment sound may enjoy a higher level of tolerance than it does within an office. The needs must be related to human activity; criteria will then result for that space or environment. This progression is seen on the lower sequence in Fig. 2.5.

'Human factors' is the study of human beings in a whole environment system. It looks at, for example, reasons for being in the system, functions and tasks, the design of jobs for various personnel, training and appraisal. A building can also be seen as a system: one that is designed to accommodate specified human activities. As some human performance aspects can be measured and quantified so can environmental performance. In the process of matching these we design and build, an activity which in its turn creates an ongoing subsystem. As data is available from both human and environmental studies a scientific basis can be constructed for its application to the production of suitable buildings: that is, the creation of a technology.

IMPLICATIONS FOR TECHNOLOGY

If the above approach to the writing of specifications is adopted then design can be an open-ended activity. The performance

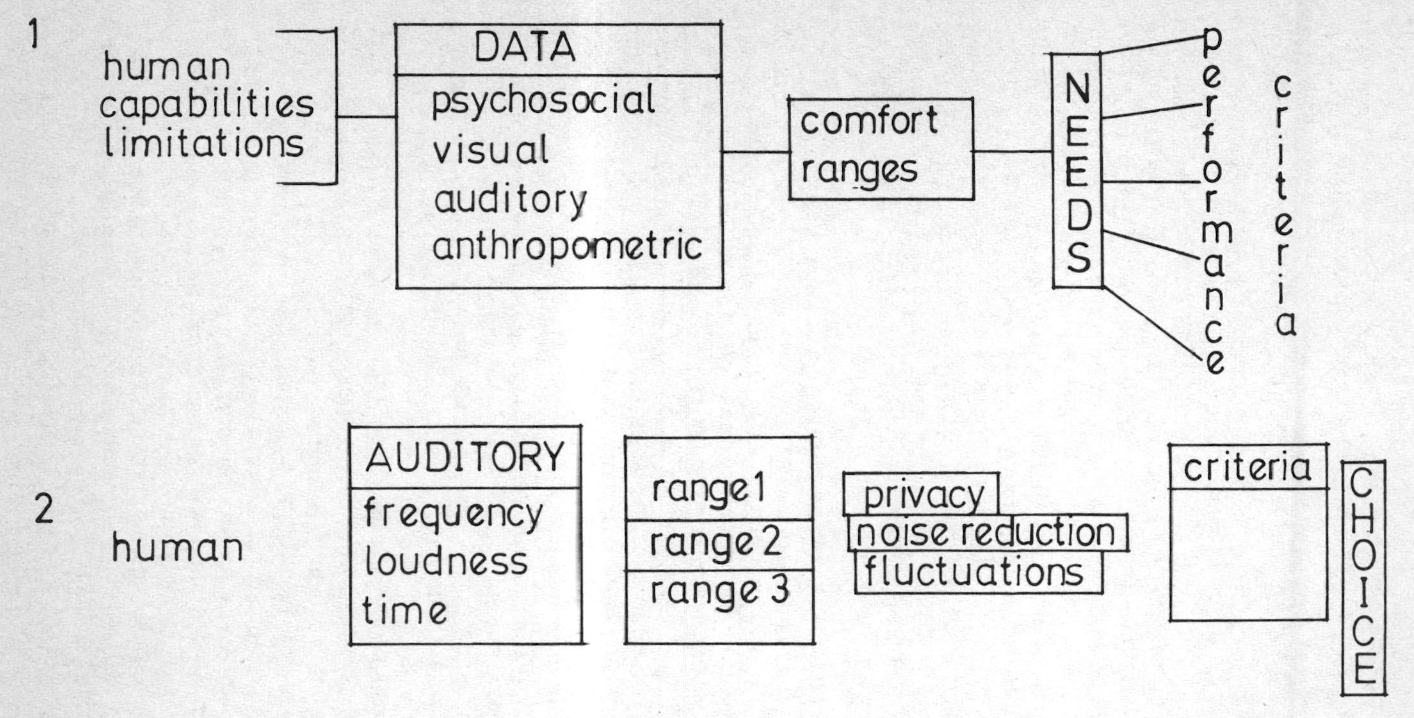

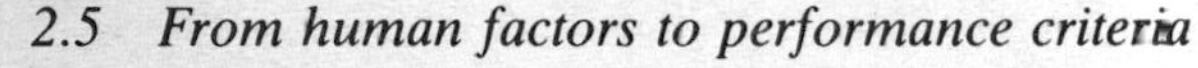

2.5 *From human factors to performance criteria*

specification sets out the conditions from which the solutions can be found. Those responsible for supplying and constructing the building have to provide the components and elements that:

(a) satisfy the performance criteria;
(b) satisfy the designer's overall concept of the building.

This means that many design decisions are taken in consultation with other parties to the construction process. An example would be the selection of windows. The performance specification sets the requirements; this information is then sent to selected manufacturers. They will provide details of their products which meet the requirements or which approximate to them. Once the choice is made the onus for compliance with the specification is upon the supplier.

Additionally, the installation will have performance criteria and whoever fixes the windows will have to ensure conformance. This could be seen, to some extent justifiably, as passing the responsibility for design from the architect to the manufacturer or installer. However, under the traditional specification, the supplier already has a duty to provide goods that are fit for purpose.

When using a performance specification the architect must be totally aware of the implications of his choices and of the need for special care. The criteria must be firmly and accurately established and adhered to; otherwise, in the event of non-compliance, the supplier is liable to remedy the fault. The architect must have a clear idea of his intentions in order to communicate them to the supplier.

The use of a performance specification can affect the technology by:

1. allowing a greater degree of control over the quality and fitness for purpose of all parts of the building;
2. giving definite criteria directly related to the designer's intentions;
3. allowing a wide variety of alternatives to be considered;
4. providing scope for change and innovation, and for designing to solve particular problems;
5. giving opportunities for manufacturers, suppliers and installers to adapt their knowledge and experience to the design;
6. putting clear responsibility on to the suppliers etc. for the performance of their contributions;
7. providing data and information to give a sound scientific base to the selection of components etc.

The following short case history illustrates the use of performance specifications and their implications in practice.

A steel-frame structure was to be built to a very tight construction programme. Continuity of site work was essential to meet the time limits. The choice of floors had to respect this need, in addition to the performance criteria laid down regarding dead loadings, fire resistance, sound resistance, accommodation of services etc. A particular performance criterion demanded by the builder was that no extra propping should be required during the construction stages. This was to enable following trades to have access to the floor below virtually immediately after the floor itself was completed. A structural screed, placed by others independently of the floor suppliers, was required to meet the need of other performance requirements; one was that the overall floor depth be as small as possible. All this information was passed to the floor beam manufacturers in the form of a performance specification. Work on site progressed well, deliveries of the precast flooring units were on schedule and being placed satisfactorily. Whilst this was proceeding, the site engineer checked the calculations of the supplier of the floors. He found that the floor would not be able to take safely the loads, due to a wet concrete structural screed being laid, even though the dead load figures were up to standard. The need to prop the beams to carry the wet concrete screed had a detrimental affect upon the whole construction process entailing:

1. a complete reorganisation of the work of the following trades;
2. the bringing in and placing of props;
3. the slowing down of the work for the floor construction;
4. restriction of activities on the lower floors;
5. extra costs attributable to lost time, extra labour and the provision of props.

The technology of the floor construction had to be altered and adapted to meet the new circumstances. The onus for the non-compliance with the performance specification was entirely upon the manufacturer of the floor beams and in consequence he was responsible for all extra costs. There was a clear line of responsibility for the achievement of concise, technological objectives.

Summary

The performance concept provides a framework within which it is possible to state the desired attributes of a material, component, service or system in order to fulfil the requirements of the intended user, without regard to the specific means to be employed. It encourages an investigation of all technological possibilities in gaining satisfactory solutions. Those with expert knowledge can be brought in at an earlier stage in design so that needs are clearly met.

It may be that these decisions of detail are carried out during the construction stage, during the builder's part of that process. An initial overall design concept is produced by the architect and all details are supplied by the specialists, fully documented with drawings and specifications etc. Such a way of working is carried out under management contractual systems such as management fee.

2. VALUE ENGINEERING

The fundamental aim of value engineering in the analysis of construction details is to provide an organised procedure for the efficient identification of unnecessary cost. It is disciplined action attuned to one specific need: that of accomplishing the functions that the customer needs and wants.

The six prime questions to be asked in the first stages of the process are related to the functional analysis of the project or a part of it. The emphasis is on technological considerations and will, hopefully, provide a basis for the optimum selection of suitable construction systems. The questions are:

What is it?
What does it do?
What is it worth?
What does it cost?
What else would work?
What does that cost?
What is it?

What is it?

In carrying out a functional analysis one must identify the function. As the concern is mainly with cost and value the search for improvement should be directed towards the potential high cost items in the project. Standard elemental cost breakdowns produced from previous projects will give an indication of these.

What does it do?

This is a key question and the answer must be centred on the main function, which is not necessarily the obvious one. The description of this function should be kept as simple as possible and an easy-to-use method has been adopted based on language grammar. Just two words are used to describe the function, a noun and a verb. Fig. 2.6 gives some typical examples of the words used. Although the words appear simple, the process of ensuring that they fully and clearly describe the function is not so simple. Care must be given to defining clearly the discrete function and to restrict it to two words. Take the description for doors. A door in a building may really depend upon the function of the opening, therefore the description would be 'provide access'. On the other hand if this door opens on to a fire escape route the discrete description, its basic function, would be 'contain fire'.

examples of descriptive words

<u>Verbs</u>

absorb	enclose	protect
amplify	filter	reduce
apply	generate	reflect
change	hold	reject
collect	improve	separate
control	increase	shield
conduct	insulate	support
create	interrupt	transmit
decrease	prevent	

<u>Nouns</u>: measurable

circuit	force	power
contamination	friction	pressure
current	heat	protection
damage	insulation	radiation
density	light	repair
energy	liquid	voltage
flow	noise	water
fluid	oxidation	weight

<u>Nouns</u>: aesthetic

appearance	effect	prestige
beauty	features	style
convenience	form	symmetry

2.6 *Value analysis: what does it do?*

The levels of analysis should be broken down further, especially when considering an element or a system within a building. It may be necessary to differentiate between the contributions of people and/or equipment. In this further breakdown one can get to the actual function of the items that make up the element or system. This will enable a pure description to be obtained which does not point to a particular solution. It will give an opportunity for a reappraisal of that item without being distracted by what is currently used or available.

What is it worth?

Value by its nature is generally subjective, although some agreement can be reached on specific data. The general aspects of value have already been discussed and in this instance the value is based on a comparative worth. Another way of looking at the worth of an item is to say that it should be at the lowest cost to fulfil its functions reliably at the right time and the right place, to the desired quality. This can be measured in monetary terms. Other values – moral, aesthetic, social, political, religious and judicial – should be fully recognised and brought into the synthesis. Aesthetics may also be given a monetary worth. An unappealing building may not fulfil its function if people are discouraged from entering as in the case, for example, of a badly-designed theatre. Since one of its aims is to attract an audience to provide the funds for the staging of plays etc., any unfilled capacity could be partly attributable to the poor aesthetics of its internal and external appearance.

Some figures given to worth can evoke an emotive response, for example, the value of a fire door. Rightly, all will be concerned that a door designed to protect life is given greater worth than one that, say, forms access to a cupboard. In this case a worth may be given which transcends the basic functional worth of the fire door.

In assessing worth care must be taken not to be influenced unduly by the possible effects of the failure of the door and its consequence for human life. The worth in this analysis in confined to function.

What does it cost?

During this stage of the functional analysis process an attempt is being made to quantify the value of the item under consideration. A value index is established as follows;

$$\text{Value index} = \frac{\text{worth}}{\text{cost}} = \frac{\text{utility}}{\text{cost}}$$

The worth figure should be kept as basic as possible so that the value index is not greater than one. The value index provides a useful indication of the premium which each alternative is costing.

What else would work?

This is the creative part of the exercise. A free-ranging approach should be adopted to produce a list of ideas. Techniques such as brainstorming are ideal for this stage. No attempt should be made to evaluate or judge until the creative session has finished. Any idea, however silly it might appear, should be recorded. Any one idea might lead to others, which in turn lead to practical alternatives. From these practical alternatives the functional analysis process is applied to produce some figures for cost and worth. These give a common measurement by which alternatives can be compared.

Value analysis procedure

Functional analysis as described is the core of the value analysis procedure. It spans three of the six stages which comprise the whole process. This is shown on Fig. 2.7. The functional analysis commences in the orientation phase and runs through to the speculation phase, with some backtracking to consider alternatives. Although presented in Fig. 2.7 as a linear process there is some cyclical feedback and adjustments.

The potential for value analysis is greatest in the earliest stages of the design/build process. Unfortunately, in the majority of cases where it has been used in construction it has not been until the stage of appointment of the builder. There is now an increasing involvement of the builder at the earlier stages of the design process, which augurs well for the fuller user of the technique. Substantial savings can be made if this method of analysis is carried out at design stage, before any hard decisions are made.

Summary

If value analysis is carried out thoroughly the result will be an optimum technological solution at the least cost. It will determine the technological detail for part or all of the building under consideration. Design will not be based on previous experience nor quoted examples. Up-to-date materials and methodologies can be considered as alternatives and compared on a equitable basis with others. Value analysis can provide a figurative measure

ORIENTATION	INFORMATION	SPECULATION	ANALYSIS	DEVELOPMENT	PRESENTATION & FOLLOW UP
what is to be studied	what is it what does it do what does it cost what is it worth	what else will do the job	what does that cost which is least expensive	will it work will it meet requirements what do I do now what is needed to implement	what is recommended select first choice and alternatives who has to o.k. it how much will it save what is needed to implement make presentation

2.7 *Value analysis: the six phases*

as a basis for comparison. It may involve some further work at the design stage but the resulting advantages should far outweigh the costs.

3. LIFE CYCLE COSTING

This concept was initially developed in the UK and known as 'costs in use', primarily by Stone (see *Building Design Evaluation: Costs in Use*). It is a technique for evaluating both the way in which a building will function and its cost throughout its life. The term 'life cycle costing' originated in the USA to describe the same idea. It is defined as 'the total cost of an asset over its operating life, including initial acquisition costs and subsequent running costs' (*Life Cycle Costing for Construction*, Flanagan and Norman, 1983).

Life cycle costing can aid design decision making in four ways:

1. by identifying the total cost commitment of the building project, rather than just assessing its initial capital costs;
2. by facilitating more effective choices in achieving a stated objective. It shows that various options will demonstrate different patterns of capital and running costs. These can be compared in like terms;
3. by acting as a management tool that details the running costs of the building and its equipment;
4. by highlighting those areas in capital and operating costs which might be reduced.

Initial costs are clear and visible during the early stages of design, but longer term costs are not, albeit they can far outweigh the initial costs. Consequently, they should have a greater influence on decisions with respect to buildings and building elements than is currently the case.

The problems facing the construction industry in adopting this approach are not slight. There is the problem of the length of time span between the phases of design and user operation; buildings are complex and the interaction between individual elements can vary, as seen in the discussions in the contextual framework chapter on Function; a change in material, element or component can mean radical changes are necessary to accommodate the variance in performance; a change in plan shape or aspect will affect initial capital costs and operating costs, such as heating and lighting. There is little doubt that the application of life cycle costing methods during the design stage will benefit the client. In the USA it is mandatory in many states to prepare a life cycle costing plan at design stage for public building projects.

Life cycle costing involves:

– identification of the overall time period;
– inclusion of all costs and revenues from the project, by time period;
– consideration of only those costs and revenues relevant to the decision under analysis;
– the effects of time, such as inflation in future years and the fact that money in the future will be worth less than money now because of reduced interest or lost income from that money.

Fig. 2.8 shows the major steps in the process. In the first step the objective must be clearly defined. Is it to provide 800 sq. m. of factory area? To choose between different floor coverings? The objectives must be unbiased and not judgemental.

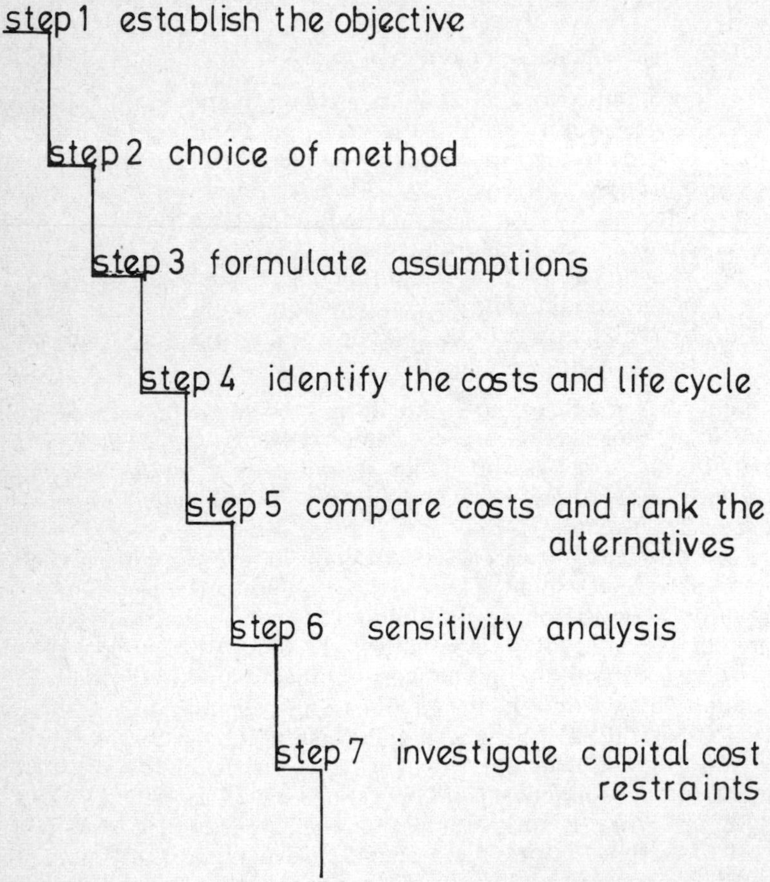

2.8 The seven steps

The second step is to determine the range of feasible methods for achieving the defined objective.

In the formulation of the assumptions, step three, it is possible to construct a full factual picture: it may be necessary to forecast such items as energy expenditure (always try to use factual data).

Step four, for each possible choice the life cycle must be determined together with the costs.

The most important step is number five; here the alternatives must be ranked, using such techniques as net present worth, savings investment ratio, internal rate of return or annual equivalent value.

For step six it may be necessary to ascertain the identification of the important variables that underlay the original assumptions.

In the final analysis, step seven, all costs must be within the budget available for the project. If this is exceeded trade-offs should be made until the optimum combination of life cycle costs has been reached.

This is not the place to go into the calculations for life cycle costing, as they may be consulted elsewhere. Here, it is the implications for design and therefore the technology of the building and its elements that are important.

The main benefit of life cycle costing is in informing the client of the full costs of alternative methods of technology. It can also offer important cost savings if implemented early enough in the design process. The later the exercise is left the more difficult will it be to alter prior decisions. There is a trade-off between capital costs and running costs and one should be considered against the other. Many examples show that of the total costs more than 50% to 60% are due to operating and maintenance, which leaves capital costs as the minor part of the calculation. In comparing one technological solution with another it has been found that a relatively high capital cost may result in lower overall operating costs, as a result of the reduced maintenance frequency of such items as services. An example is the fitting of direct control valves to a heating system, which monitor all the spaces which the unit serves. The initial costs will be greater owing to the large number of individual controls required to serve each space. One overall control would be much cheaper. But in the operating of the system it will not be possible to manipulate closely the heat requirements in each space, so excess heat may be used, leading to unnecessary cost. A boiler with a larger capacity than required may also be more cost-effective over a long period as it will not be constantly working to maintain the desired temperatures; wear and tear will be reduced, thereby reducing maintenance and lengthening the life of the boiler.

In the selection of appropriate technology, whether at design stage or building stage, it is important to understand not only its functional performance criteria but also its monetary performance criteria. A technological solution taken just on its ability to function at any cost is, in this time of diminishing resources and high costs of money, irresponsible. The construction industry must provide society with adequate functional buildings which do not drain future resources. If more than necessary is spent on operating and maintaining a building it leaves little for further investment in other commercial enterprises or new buildings. Life cycle costing techniques will go a long way towards reducing future costs and it is an exercise which should be an integral part of any process.

Life cycle costing techniques:

- focus attention upon the relationship between capital cost and running cost of the building and building components;
- provide a methodology and framework to enable the design team to estimate the total cost of what would happen, what should happen and what will happen;
- provide a checklist of occupancy cost items to encourage the design team to bridge the gap between the design and construction phase and the occupancy phase of the building's life;
- use information contained in the bill of quantities and on drawings and specifications as a basis for a life cycle cost management system for new or existing buildings. This will identify those areas in which the running costs might be reduced, either by a change in operating practice or by replacing the relevant system. It is also important to the design team that there is a feedback on the cost and performance of the building in use;
- structure the data in a hierarchy of levels, thus allowing a more co-ordinated approach to capturing cost information and performance data for both new and existing buildings;
- evaluate the total costs of design options in a simple manner.

Building clients have shown that they realise the importance of knowing the operating costs of their buildings. This concern is expressed in the manual on their system for designing and constructing buildings (*The British Property Federation System for Building Design and Construction*, 1983). They call for a master plan to be produced during the early stages of design: 'At all times (this) should provide the best possible estimate of the final cost of the project, of the future cash flow and of the future costs of the building'.

Summary

Life cycle costing is primarily an accounting procedure applied to the decisions made regarding the technology to be adopted. In the testing of these decisions against future costs it may be that the choice is altered from that first proposed. A cheaper initial solution may have the greater long-term costs: in that case a more expensive solution, with its subsequent technological ramifications, would be adopted in order to reduce future costs.

4. ENERGY ACCOUNTING

Since the early 1970s there has been considerable concern about the use of energy. A massive price increase, together with the realisation that most fossil fuel sources are finite, brought into focus the need to reduce and conserve energy as much as possible. In the UK there has been only limited government intervention. The policy has been to raise the price of energy consumed by industry, commerce and domestic users to levels that are above those required to sustain production and investment costs. Since 1974 there has been some further legislation regarding the allowable heat loss through the external elements of a building. The 'U' values have been improved considerably, with the possibility of further improvements for new buildings. Also, nearly all classes of buildings are now embraced in the legislation. However, there have been no statutory limits placed on the amount of energy that can be used. In other words, so long as a building is shown not to lose heat generated within it there is no control over the amount of heat that can be produced. A guide has been produced by the Chartered Institution of Building Services Engineers, *The Building Energy Code*, Volumes 1, 2, 3 and 4, which gives designers the ways and means of calculating a building's energy use. The Royal Institute of British Architects has also considered that energy conservation is a vital ingredient of good design. This interest is reflected in the provision of an energy desk and sub-committees considering relevant issues and giving advice. The British Standards Institution has also prepared a Code of Practice for designing with energy conservation in mind.

Research in this field is now extensive and continuous, but there is still much to discover and to implement. There are now some basic findings which can be applied to general building design, which will ultimately affect the technology. Those considered below are taken from the CIBS Building Energy Code.

(a) Site Factors

Geographical

(i) climate – the aim is to reduce unwanted heat gains during summer and heat losses during winter.

(ii) exposure – unnecessarily high buildings should be avoided as direction and strength of wind can materially affect the energy balance of a building. Existing topographical cover should be used where possible.

(iii) fuel – in the UK the sources of fuel are comparable in availability and cost.

Local

(i) pollution – from such as noise and traffic fumes. If this is to be avoided then sensitive areas should be furthest away from the site boundaries.

(ii) shape of site – this may dictate a building uneconomic in energy terms. May require reappraisal of requirements.

(iii) surrounding buildings – may reduce daylight and useful solar gain. It may be necessary to increase energy requirements in these cases. The new building, in place, may create unacceptable wind patterns with corresponding increases in ventilation and transmission losses.

(iv) access and vehicle parking – owing to access road and parking demands the orientation of the building may be affected causing an adverse affect on its energy requirements.

(b) Building Form

(i) shape – if the ratio of envelope surface area to usable floor area is kept to a minimum then this aids energy conservation through the building fabric. A circular or square plan is preferable. For a given treated floor area the shape of a building should be nearly cubic if heated. If air-conditioned a lower rise building is preferable. The less glass in the windows the nearer the optimum shape of the air-conditioned building is to a cube.

Buildings large in floor area which produces a low ratio to external surface area are more economical in energy usage terms, but benefits of natural lighting and ventilation may be lost.

Deep plan structures may use more energy than relatively narrow ones as they will need artificial lighting and air conditioning.

(ii) mass – the most relevant elements are floor slabs, external wall and roof, with most mass being concentrated in the intermediate floors.

Intermittently heated heavyweight buildings store so much heat in their structure that the benefits of on – off heating are lost. Lightweight structures are better when intermittently heated. The risk of condensation is higher in lightweight structures.

The roof's mass is more significant than the external walls, especially as the number of storeys diminishes. In a single storey building it is comparable with the floor slab. A heavier roof is sometimes desirable to reduce the influence of solar gain. The mass of the building must not be isolated: the optimum 'U' values of walls and roof are related to the mass of the structure.

A well insulated lightweight building will show thermal characteristics similar to a less well insulated heavy mass structure.

(iii) orientation – has little or no effect on heated buildings, although south-facing windows may benefit. If air-conditioned, the major axis should be east-west; the greater the glass area the greater the benefit.

(iv) roof insulation – a well insulated roof is more important on buildings less than four storeys high than on higher buildings.

(v) wall insulation – diminishing returns are gained by improving the 'U' value of walls. In a heated building heat loss through the walls could account for 50% of the total loss.

(vi) floor insulation – any floor exposed to the external environment should be insulated.

(vii) glazing – a tentative suggestion is that between 20 and 30% of the outside facade should be glazed, giving some natural lighting and ventilation to a depth of 6m from the windows. This assumes that the external environment allows the windows to be opened.

There is still little evidence of economic advantage for double glazing other than in buildings continuously heated.

(viii) lighting – the optimal standards for all illumination depends upon the tasks to be performed in the building. Lower standards for shallow buildings can be employed, as some advantage can be taken of natural illumination.

(c) Thermal Insulation

Thermal insulation can be used in two ways to reduce the building's energy requirements:

(i) building envelope – consideration needs to be given to the area of glazing and its effect on artificial lighting and wintertime solar gains; the thermal inertia of a building; the method of operation and control; fuel type and cost etc. when calculating the required degree of insulation.

(ii) heating, ventilating and air conditioning services – should be thermally insulated when used to transport, store or generate fluids or gases at temperatures different from their surroundings, except where the transfer of energy is part of the service function.

Demands and targets

In order to be sure of the energy used in a building it must be calculated. This total energy requirement is called a 'demand' and is related to that particular building. Empirical studies and calculations have been developed for various building classes, such as shops, factories, warehouses, residential, hotels and educational. Typical values have been assigned to each class, deemed to be those required to meet national needs regarding the use of energy. It is possible, therefore, that the values may change according to changes in energy availability. Abundant and cheap energy could result in a relaxation in these values. They are applied to the building and related to areas of walls, floors, number of storeys, and floor-to-ceiling heights. In this manner a 'demand target' is formulated. In both the calculation of the demand and the demand target there is a separation between thermal demand and electrical demand. This gives some flexibility to the designer. For example, if less glass is used in the envelope the thermal demand will diminish (less heat loss) but there will be less benefit from natural daylight and any useful solar gain will be reduced.

This will give an increase in electrical demand because of the need to use more electricity. By improving the 'U' value of the window wall and, say, double or triple glazing the windows the same window area can be maintained. This will then reduce the electrical demand by using natural daylight. The total energy demand will be less, but at a capital cost.

In order to ascertain the economics of this decision, life cycle costing techniques need to be used. Any technological proposal, or its alternative, should be cost-effective over the life of the building. It may be that where alternative demands meet targets in every case then the most cost-effective solution should be adopted. This analysis should be carried out at design stage before any firm decisions are made.

In order to calculate the demand or demand targets information about the following is necessary:

building type;
dimensions;
partitions;
windows – as a percentage of the facade;
orientation;
thermal transmittance 'U' values;
thermal admittance 'U' values;
location;
occupied period;
design conditions – internal temperature air change rate;
occupancy;
design illuminance;
proposed building services – heating system, control, hot water system, boilers, lighting.

Using data in the Building Energy Code the demand and demand targets can be calculated. The final analysis is expressed as a ratio of Demand to Target, e.g. 0.9 is a demand lower than target, which is beneficial. Where the thermal or electrical demand ratios are in excess of 1.0, but the average total is less than 1.0, then this is acceptable. It is important that the total demand does not exceed the demand target. If it does then remedial work will need to be done to reduce the demand. This may involve reducing the number of storeys, increasing thermal insulation, using more efficient energy sources etc. What it implies is that the technological solutions first proposed may need to be altered or changed.

With larger and more complex buildings, with sophisticated services and controls, it is necessary to ascertain the energy use at an early stage. Clients and users are demanding that buildings should run economically; energy is a major proportion of their total bill and means to reduce its consumption should form an essential element of the design process.

Summary

With the publication of the BSI Code of Practice on energy effectiveness in buildings, the attention of all parties to the construction process is drawn to the need to consider the use of energy. Technology has to take full account of the need for energy conservation in the erection and in the materials used for new buildings or for refurbishment. It is a concern of both builder and architect. With the use of such aids as the Building Energy Code it is possible to quantify and compare various solutions and to demonstrate the levels of energy that will be used throughout the life of a building.

5. FEEDBACK: HUMAN FACTORS TO PERFORMANCE

So far four identifiable techniques have been considered which are becoming essential ingredients of the design mix. The architect as the creator of space is the paramount factor in the design process, as previously discussed under the headings of human factors, value analysis, performance specifications, life cycle costing and energy demands. Each has been promoted in its own right as a factor of design. Ideally, all will become part of the design process. In many ways they overlap in their concerns, the prime objective being the achievement of the optimum solution. This approach to design is based on the use of data and its relationship to the proposed building form.

It is appropriate at this point to present a case history which shows that one cannot rely solely on information presented as authoritative, nor ignore the actions of the people who use buildings. The human factors approach to design is not just concerned with the ultimate goal of producing a building: it is also concerned with the system of the design process and the reactions of people when caught in unfamiliar circumstances (as in the stand fire at Bradford City Football Club in 1985). It is *human* factors and inter-relationships, failures, neglect or lack of knowledge that, when woven into the design, create flaws and mistakes. It is correct to give due attention to the process of design as well as to its product. The following account illustrates how human failures can be prime contributors to disasters; as an editorial in the *Architects Journal* (29 May 1974) said soon afterwards, the errors were 'so ordinary and casual that many of them could have occurred in any architect's office. Who has not specified a new material – even on occasion a whole string of untried components – without being fully aware of all the problems that are likely to ensue?'

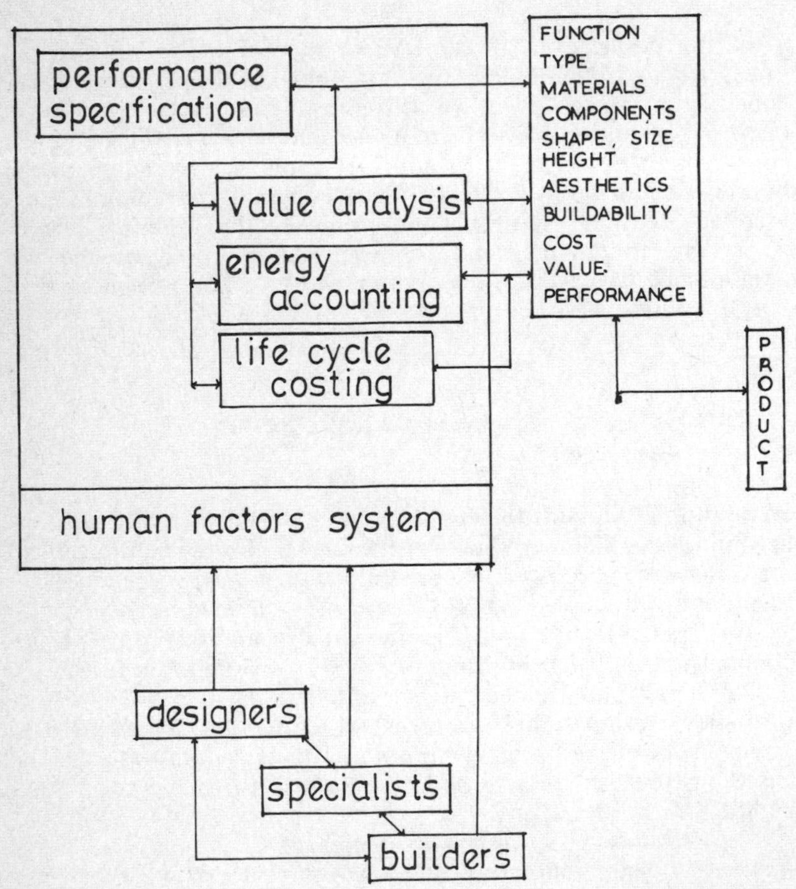

2.9 *Interrelationship of the five techniques to the product*

THE SUMMERLAND FIRE

Fifty men, women and children died in the fire at the Summerland Leisure Centre on the Isle of Man on 2 August 1973.

The facilities offered in the complex were for entertainment, music, singing, dancing, eating, drinking, sunbathing, bingo, pintable games and table tennis. It had a capacity of about 5,000 people and was a big attraction on the island.

Two architects in private practice were appointed as principals for the project; they brought in, to the approval of Douglas Corporation, the client, associate architects with established experience in designing leisure centres and enclosed shopping precincts. Overall responsibility, however, still lay with one of the two principals.

The two-acre site was constrained by the promenade and a rocky cliff. Access to the site was difficult. Adjacent to the leisure centre and already built was the Aquadrome, a swimming pool. The building was on seven levels (the main entrance being at level 4, the equivalent of the ground floor): level 1 (discotheque), level 2 (lower downstairs, children's play area), level 3 (large balcony, upper downstairs), level 4 (solarium), level 5 (marquee, show bar, terrace and pool bar), level 6 (leisure terrace), and level 7 (cruise deck terrace). The two lower floors were of reinforced concrete and those above level 4 of large steel U sections. The final structure of the terraces was steel supporting wood floors. Although there were staircases throughout the building, only two were enclosed. Most were in reinforced concrete, but the main flying stair rising from level 4 to levels 5 and 6 was a completely open stairway with hardwood open treads. A wooden stair, added at a very late date, led down from level 5 to level 4.

The main south wall was formed from Oroglas, an acrylic dome square panel with diamond profile, as was the entire roof. There had been a close association between the architects and the manufacturer of Oroglas as the architect had contributed to a symposium on this type of glazing for space structures. During a discussion a Ministry Research scientist had asked, 'is there a problem when fire causes the destruction of the complete structure?' The reply to this question and another – 'Have any fire tests been carried out on a completed structure?' – was not given in the subsequent publicity leaflet. The associate architect did not question its use in walls and roof.

The eastern elevation was clad in coated corrugated steel, colour Galbestos, supported on steel angle rails in turn supported on the main structural steelwork.

Even though this was a publicly sponsored building (subsequently let to a leisure company) it still had to comply with the Isle of Man by-laws. The Local Government Board had the right to 'suspend, alter or relax or dispense with compliance thereof'. There were three significant by-laws in the context of the later fire. One said that the external walls of any building should be non-combustible through-out and have a fire resistance of two hours. Another stated that the roof of every public building should be so covered as to afford adequate protection against the spread of fire into the building or to adjoining buildings. The third said that the cavity between leaves of a wall formed of or containing combustible material should be fire stopped at the junction of the wall with any other wall or with any floor, ceiling or roof, and at intervals of not more than 15 feet. Subsequently three applications were made for approval. The principal architect stated in a letter to the Borough Engineer that 'the enveloping structure is, in fact, an acrylic glazed space frame, no part of which is combustible, but both the acrylic sheets and the alloy framing cannot be regarded as fire resistant'. The chief fire officer made it clear that Oroglas afforded no fire resistance, concluding, 'Since the complex does not present an exposure hazard to any other building, there is unlikely to be any interference with means of escape. I raise no objection to the suggested construction.' A waiver was given to suspend the whole of the first mentioned by-law. The same suspension was given with respect to the Galbestos wall. The building required a theatre licence, but just after the opening the chief fire officer wrote,

> There is a good deal of work still to be done before it can be said that all safety requirements have been met. However urgent steps are being taken to ensure completion, and in order that the opening of the complex should be legalised, I recommend that the certificate of fitness be issued now. I recommend that this be accompanied by a letter making it clear that its issue is conditional on all safety requirements being completed without delay.

Internal fitting out work used the combustible materials of plywood on softwood studding, exposed timber and plywood, Decalin wall linings, decorative ceilings, a hardwood and perspex sliding partition and a polythene curtain.

The building was successful, contributing 13% of the island's tourist revenue in the first full season of 1972.

The fire started in a dismantled section of a fibreglass kiosk, leaning close to the Galbestos wall. It was set alight by three boys and soon blazed fiercely, collapsing against the wall, which by this time was aflame on its outside coating. The sheet metal of the Galbestos conducted heat quickly to the internal cavity, creating inflammable vapours from the inside coating. These ignited, creating fire between the internal fibreboard lining and the external sheeting. The fibreboard ignited and a massive spread of fire occurred at level 4. It spread upwards to levels 5, 6 and 7 via the front edges of the terraces, in voids between the Galbestos wall and the interior partitioning and through a 'chimney' between the south Oroglas wall and the ceiling/floor edges. The massive fire quickly spread over the full height of the building above level 4, but did not penetrate below the reinforced concrete floor of level 4 into the lower levels. Everything combustible at the upper levels was eventually destroyed.

Staff tried unsuccessfully to extinguish the fire. The fire alarm was set off but did not sound in the leisure centre, only at the fire station. The main electricity supply was switched off, but the automatic generator did not start. An enclosed staircase was plunged into darkness. When the fire brigade arrived the chief fire officer decided Summerland was unsavable, but that fire could be prevented from spreading to the Aquadrome. This was successfully achieved. However, smoke, followed quickly by flames, rushed through the Summerland building. Within a few minutes people who had not found an unblocked exit were dead and severe injuries were sustained by others who fell over in trying to reach unblocked exits. It was an inferno.

After the fire a chorus of blame arose. There were disputes over the fire source; over the allowance of Oroglas and Galbestos on the structure; over who was responsible. An official commission was set up to discover the causes of the catastrophe.

The main factors in the disaster, as identified by the commission, were as follows.

The rapid spread of fire was caused in part by the design and construction of the building and in part by the failure of the Summerland staff to take prompt and appropriate action. The building might have been saved if the fire brigade had been called sooner.

If Galbestos had not been substituted for the reinforced concrete initially planned for the eastern wall the disaster would not have occurred. In granting waivers, compensatory action should be taken to reach the same levels of safety obtained before the substitution. The inflammable coatings of Galbestos accelerated the fire, made worse by the internal void. Furthermore Decalin was used instead of plasterboard as material for an internal wall. The commission thought that this error might have been the biggest single structural contribution to the disastrous consequences of the fire. The designer had first seen Decalin the day prior to specification; he did not know the properties of the material, nor whether or not it was combustible. Fire stopping would have delayed the fire, but the void was not effectively stopped.

It was inevitable that fire would spread throughout the building, owing to the voids in the structure and the openness of the internal layout, and the fact that the roof was of a material, Oroglas, that would fail immediately and vent the fire. There was no compartmentation by means which a defensive fire spread system could be introduced. One way of protecting this kind of building would have been a sprinkler system.

The architects thought that Galbestos was virtually incombustible. No questions were put to the manufacturers, whose advertising literature was unclear. In addition, the manufacturers of Oroglas did not realise that the material burnt so quickly: the roof burned out in ten minutes.

The high number of casualities was seen to be caused by two factors: the extremely rapid development of the fire and the delay in evacuating the building. Faults in management and in design and construction contributed to the defective evacuation. The stairs as means of escape were inadequate. The exits did not allow for easy egress from the building. No procedures for unlocking doors had been instituted, and some fire doors were padlocked (the fire service had complained to the management about this). No proper signs guided the way to the exits. Some doors leading on to stairs meant as fire escapes were not self-closing or fire-resistant. An authorised opening had been made in a wall which allowed smoke into a staircase.

The conclusions of the commission were as follows:

1. No efficient design control had been imposed by the architect.
2. There had been misunderstandings between Douglas Corporation and the Local Government Board in the matter of giving waivers.
3. Tests on materials had been limited and further applications could not be safety predicted.
4. The nature and character of official inspections of work in progress had not been clearly understood.
5. The tenants of Summerland had never seemed to be aware of how vulnerable it was or might become. (But their insurers had said that if a sprinkler system was installed a very substantial reduction in premium would be offered.)
6. The waiver of a by-law is only justifiable when no reduction in safety occurs. Although the Oroglas was not ignited until fire came from within the building, and was not the prime cause of the fire's spread, the combustibility of Galbestos, together with the fibreboard linings and the floor to ceiling relationship, was lethal.
7. The problems of fire fighting should have been considered by the chief fire officer when first receiving the plans for submission. He was later not consulted on materials and means of escape.

In the commission's final judgement, there had been no villains, just human errors and failures, poor communications, and too much reliance on an 'old boy' network.

The following recommendations were put forward by the commission.

1. *Architects*

(a) A named person should be in charge from the outset in the design of a building, and take the major design decisions.

(b) Architects and clients should carefully consider the requirements and performance of a building in use at the stage when conceptual designs are produced and before proceeding with the details of design and the later submission of plans to the relevant authorities. Given this collaboration and advice, designers must take responsibility for agreed decisions.

(c) Architectural training should include a much more extended study of fire protection and precautions.

(d) Full provision of escape and fire protection means should be included in the earliest designs adequate to the maximum foreseeable occupancy.

(e) With large public buildings used for assembly and entertainment, and containing a lot of necessary inflammable material (notwithstanding any proposal for retardant treatment), the design should include a sprinkler system.

(f) Voids with combustible interior surfaces should not be unnecessarily incorporated in public assembly buildings. If functionally necessary they should be provided with permanent and reliable fire stopping, or with sprinklers.

(g) The difference between building materials and structures in use and those tested in standard situations must be realised. If necessary carry out special investigations.

(h) Instability to steel sheet claddings may result from thermal and vibratory movements and consequently affect their joints with other materials.

(i) Planning should include a schedule of means of escape. If functions decree the separation of parents and children then such separation should be confined to one level. If parents and children are using different floor levels generous means of escape must be provided, together with arrangements for marshalling.

2. *Building inspectors*

(a) Inspections during building should be conducted formally and precisely, both by architect and by local authority inspectors. They should be recorded, to confirm that the building is being built in accordance with the approved plans and the relevant by-laws and regulations.

(b) No public buildings should be occupied until a satisfactory official inspection has been carried out and a completion certificate issued.

3. *Public authorities and entertainment or assembly hall managers*

(a) Detailed and up-to-date plans should be available in all occupied buildings.

(b) The Manx building by-laws and theatre regulations should be revised.

(c) Certificates under the theatre regulations should specify any exemption granted and any special conditions imposed.

(d) Before granting an application for waivers of a by-law or regulation an authority should satisfy itself that the standards of safety for occupants as set by law or regulation will be maintained by some other means.

(e) Full provision for protection and means of escape in case of fire should be integrated into the design at the earliest stage.

(f) Fire routines should be reviewed regularly, checked and practised, with full staff training.

(g) Inspections should be carried out whilst members of the public are present.

(h) If the required standard cannot be met after fire fighting instruction to a fire fighting group, the fire brigade to be called to every alarm.

(i) Adequate access to the building for fire fighting equipment to be maintained.

(j) Emergency lighting equipment to be tested regularly and a record kept of such tests.

(k) Diagrammatic plans showing escape routes should be publicly displayed, together with prominent signs showing the routes to the emergency exits.

(l) Emergency doors never to be locked whilst the public are on the premises, even if keys are available in adjacent boxes. All exit doors to be readily openable from within, with suitable fastenings on advice, if necessary, from the fire authority.

4. *Manufacturers*

(a) If manufacturers are expected to take responsibility for some part of the building's performance these should be clearly defined and agreed, with the client's knowledge.

(b) The fullest information possible should be provided about the fire properties of building materials.

In respect of acrylic sheeting the following was recommended;

(a) Until more information is available its use should be confined to situations in which the hazard it might present is minimal.

(b) it should not be placed within 3.5m of any point which would normally be within the reach of persons inside or outside the building.

(c) It should not be placed within 6m of any combustible material, or any point where combustible material may be put.

(d) Its use should be limited in multi-storey structures. Each area of acrylic roofing or cladding should be sufficiently separated by non combustible material.

(e) External exposure hazard to any occupied building roofed or clad with acrylic sheeting should be avoided by the provision of adequate separation distances between the building and its site boundaries as laid down by modern regulations or codes of practice.

(f) If recommendation 4(e) is not followed the roof and/or cladding should be protected by a reliable and effective water spray system.

(g) Higher standards of means of escape should be provided when a building roofed or clad in acrylic is intended to be occupied by more than one person to every 3m sq. of the net area.

(h) When acrylic sheets are used for roofing or cladding, the edges should always be protected against ignition unless the risk is otherwise obviated.

5. *Other general recommendations*

(a) Fire alarm system manufacturers should investigate the possibility of designing systems to sound when the wiring is affected by fire, rather then letting fire render the system inoperative.

(b) Emergency generator sets fitted with cut-out switches should also be provided, with either an interlock or a buzzer or a warning light to prevent the set being inadvertently left inoperative when unattended.

(c) Where a fire alarm system incorporates a mechanism whereby either the public warning signal or a call to a fire brigade can be placed on a delay, the mechanism for this purpose should, if possible, be so designed that the period of delay cannot readily be altered without reference to the fire authority.

Summary

The recommendations arising from this incident have made an impact on approval procedures and fire technology for buildings. There is now firmer collaboration between designers and fire officers. New materials and technologies are being submitted to stringent tests to ascertain their characteristics in fires. Designers have been alerted to the need to ensure they are familiar with fire technology. Building owners and users are taking greater responsibility for carrying out adequate procedures for preventing the spread of fire and providing adequate means of escape.

How does this case study relate to aspects of design, particularly human factors? The main point to be made is that although there may be procedures, experts, materials, techniques, etc. it is the human beings carrying out a process which can affect its final technological outcome. The process of design is just as important as meeting the requirements of the client's brief. Also, the building must be perceived in its complete entirety, in use: that is, when occupied by people. There must also be a recognition that failure may occur. In the case history the risk of fire was probable, whatever the materials used, either combustible or non-combustible. The behaviour of people in that situation must be taken into consideration. In Summerland, a building with occasional users, it should be appreciated that people will not know its layout. The design can be criticised as not catering adequately for the user. In any emergency evacuation would have been difficult.

In the human factors approach to design the designer's solution should take account of the way a building is to be used and how it can cope with an emergency. It will be extremely difficult to make a building (or any part, component or equipment) fail-safe, but there are known risks which can be alleviated.

It would appear that the concept of performance specifications was not utilised in the design process for Summerland. Materials and techniques were chosen on aesthetic grounds rather than to ensure that performance functions were met. If the performance criteria had laid down in detail the requirements for sound, thermal and fire resistance for the cladding it is likely that a far more suitable material would have been forthcoming.

If an approach had been adopted which incorporated value engineering procedures the problems of the building's performance might have been highlighted. Value engineering analysis can be applied to spatial relationships, e.g. escape routes, as well as to component and element constructions.

It is unlikely that the designers considered the building's use of energy. Having very large open areas is not conducive to minimising energy consumption. Walls with high thermal transmittance materials will accelerate energy use.

Whether life cycle costing techniques would have affected the adopted solution is doubtful. It might have shown that running costs would be relatively high, but this was before the time of very high energy prices. The external wall constructions of Oroglas and Galbestos were thought to require little maintenance, so on that basis it could be considered suitable.

Many instructive lessons can be drawn from the Summerland disaster which are pertinent to building design and subsequent construction. For example, the builder was not held to be involved in causing the fire. On the other hand, could the builder make a contribution to the design process which would avert some problems? Is it the owner's or the designer's responsibility to ensure adequate means of escape? Who should be responsible for appraising the design? Should one person be responsible overall for decisions relating to the building's design and construction?

QUESTIONS

1. Should the technique of value analysis be applied to all building work and all parts of buildings?
2. Describe, using a particular building element or component, how energy conservation requirements can determine design technology.
3. Discuss whether or not the techniques of life cycle costing should be used for most building structures.
4. To what extent is it possible, or appropriate, for designers and builders to be involved in influencing the design in the direction of energy saving? To what extent is it the responsibility of services engineers?
5. How can the concept of human factors analysis be incorporated into the design process? What are the difficulties and how effective could it be?
6. With reference to the Summerland fire, is it realistic to expect that all risks can be ascertained and reduced to a minimum?
7. If the builder is involved at an earlier stage in the design process, outline how he could influence the choice of materials and components if working with a performance specification approach.

4. Design and Production

Integration of the design and construction processes has been advocated for many years, but there is little progress towards its resolution. Bowley (*The British Building Industry*: *Four studies in response to change*, 1966) sums it up thus:

> This problem has been referred to so often and it is such a burning issue in the building industry today that it is important to try and clarify the main points. It will be convenient to start from the product, the building, and ask what are the most appropriate methods of organising the various jobs that have to be carried out in order that the building may ultimately be erected. This means starting by ignoring the individuals who customarily perform particular functions and instead consider how these functions should be organised. Clearly there are three main functions: the design of the building; making the decision as to whether or not to undertake to produce it on any particular terms; and the organisation of its production if it is decided to produce it.
>
> The design function as such includes design in all its aspects, that is design of the structure required by the customer, the choice of materials and the provision of services and amenities, as well as the overall arrangement of space, elevation and layout on the site. Enough has been said in earlier chapters to show that unless these aspects are taken into consideration from the start and the design worked out as a unified process on the basis of all expert knowledge relevant, a building may be less efficient than it should even though the building owner may be unaware of this. It has also been made clear in earlier chapters that if costs are of any relevance, except as a rather indeterminate ceiling, it is of major importance that the production problems involved by particular designs, and their effects on costs, should be considered from the start. This is necessary so that the cost of deciding things one way rather than another should be taken into account at all stages.

The issue still burns! There have been attempts to integrate design and production, but not tamper with the individual roles and training of the industry's protagonists. The vested interests of the various professionals seek to retain the status quo, or endeavour to convince that each is best suited to control the design/production process. Hence, there is a whole array of procurement methods and their management, viz. project managers, project management, fee management, alternative method of management, BPF system of management, etc. All tinker with the structure of the design-production process: none offer to restructure. Perhaps this is too much to expect. Any progress is piecemeal. The point of change is to ensure that the right technology is used at the right time, to the correct cost, in the shortest possible contract time, so that a worthwhile building is produced. It might be said that the biggest influence for change will arise from the initiative taken by the major clients as represented by the British Property Federation. In the final analysis it is the clients who pay and whilst it is not advocated here that monetary policy should rule all decisions to the detriment of quality, performance etc., heed must be taken of the paymasters. Building is a service industry, and the service offered is a product, generally of a one-off type. The client is concerned about the process of construction as it determines the time for completion, the quality levels and the costs. This makes it different from many other products, where the emphasis is primarily on the end result. The public, as customers, have little concern with the manufacturing process of cars. On the other hand, customers for building work have a major concern with the process as well as the final result. If any building work is carried out to premises under his care the customer involves himself in its progress as of right. The BPF have legitimated this concern and expressed it in the publication of its manual.

A current term describing the concern of the integration of design and production in a technological sense is 'buildability'. In essence it means the ease and efficiency with which an element can be put together. To take a simple example, the designer could state the width of brick piers between windows to be 825mm as this suits the internal layout. To construct this would involve considerable labour in setting out and cutting bricks to size, creating waste. With a minor modification to the internal layout the pier dimension could be made to match the brick sizes. The external appearance would hardly be changed, but if there were many such situations on a building the savings would be considerable. A higher degree of buildability would have been achieved. Although this factor should be resolved by the designer a further discussion of buildability will be included in the chapter on Production.

To return to the roles of the professional in the industry: their perceived functions have an effect on technology, as in the situation of the Summerland Leisure Centre, where the architect determined the technology. As buildings become more complex in function and services it is increasingly unrealistic to expect one person to have enough knowledge to cope with every technological decision. If the architect had drawn on advice other than that supplied by the manufacturers; the inherent dangers of the materials used at Summerland could have been foreseen. This means that in practice more people need to be brought in at the design stage. As a consequence the role of the architect may diminish.

In the BPF proposals the project leader is responsible for the whole project from inception to handover and possible running. He or she is not deemed to come from any one profession or hold a specific qualification. The ability to perform the function will depend on the fitness of the individual. The abilities required must be appropriate to successfully co-ordinating all contributors to the building process. The project manager could be an architect, engineer, builder, quantity surveyor or services engineer.

To consider this from another direction: the builder could be brought in at an early stage of the design. It may be that this will mean the builder will not carry out the actual work on the site: he may just be consulted over the buildability aspects and costs and construction methods. He will be giving technological advice.

No doubt a further shift in roles and relationships, together with the establishment of new roles in order to cope with modern requirements, will be seen. There is a need to cope not only with rapid changes and advances in technology but also with the indirect processes which these innovations bring in their wake, such as information technology and computers.

QUESTIONS

1. What are the implications for the technology of construction if the project leader is a builder?
2. Should we be more concerned about the best methods of construction than with arguing about who should do it?
3. Design/build organisations are one means of achieving integration between production and design. Discuss the effects they have on the choice of technologies available to the client.
4. The Alternative Method of Management installs the architect on the site as a manager, who designs and organises the work which is carried out by specialist contractors. Discuss the ramifications this may have on site technologies.

5. Computers in Design

There is no escape from computers. They have become an indirect and integral part of everyday life. Personal finances are recorded and dealt with by computers; national statistics are stored in computers; economic forecasts and policies are guided by computers. Directly computers are playing a large part in the design of buildings.

The computer with its software is a sophisticated tool, capable of carrying out tedious, menial or complex tasks quickly and accurately. As yet, it cannot act as a substitute for a human being, but Expert Systems programs may be developed which can evaluate the multiplicity of factors to be considered in the design of a building. Many options can be tested and results compared. From these results preferred solutions can be produced. However, it is the human being who must take the final decision. Not far into the future it should be possible to set up instructions from the brief and, all alternatives having been considered, the computer would produce the optimum design solution; this would then be broken down into elements and components. The computer would produce programs for parts of the building to enable robotic manufacturing processes to produce the goods. A consequential and times programme could be produced and the parts manufactured and delivered to site when required. Machines on site could also be computer controlled (from the master program) and the construction of the building would be automatic. Far-fetched? Robotic manufacturing processes already exist. There are computers and software (computer-aided design) that will produce drawings and information from an overall design.

Sequential programmes are commonly produced, together with materials quantities and call-off schedules for co-ordinated deliveries. Expenditure and costs can be forecast and subsequently monitored by computer. Structural design can be undertaken by computers, together with techniques of value analysis, life cycle costing and energy demands: the building can be readily appraised. In essence the constituent parts of a master program are available which could carry out the design and build process. When the building is complete a computer program could run and monitor the performance; this as an offspring from the master program. (The computer control of building environments is a reality.) The type of program described above is known as a knowledge-based or expert system. In the medical profession there are programs which are able to diagnose patients' illnesses, recommend sequences of actions and tests and advise on suitable treatment. The expert in a particular field of medicine has put knowledge into a computer so that those not having that specialism can 'tap' an information source easily and quickly without having to consult the specialist in person.

It is feasible that an expert system could be created which will, say, design a flat roof for a particular project. This will consider the desired performance criteria such as span, strength, durability and integrity, with structure, thermal and sound insulation, weathering, fire resistance, cost, maintenance aspects, possible failures, production sequence, and so on. Incorporated into the system will be the experience and knowledge of many practitioners and, hopefully, the mistakes of the past in flat roof design and construction will be avoided. The program will be so constructed that questions can be put to establish the reasons for the solution offered. The computer's reasoning can be investigated. When a suitable roof construction has been designed, drawings can be produced directly from the computer, with full notation and specification, together with quantities and cost. Ordering schedules can also be issued which take into account the manner in which the materials are supplied by the manufacturer. Guidance notes can be given for the successful construction of the roof.

The desirability of this type of program can be debated. It can also be seen that to collect and to synthesise the knowledge relating to flat roofs, let alone any other construction elements, is a major and expensive task. One would also need to update continuously the information to take account of new materials, legislation, knowledge etc.

What is worth considering is the use of the computer as a means of communication between all the parties in the construction process. Mention has already been made of the divide between design, production and performance. There is still little movement towards the breaking down of barriers between the professions, and compared to other industrialised nations the UK's building industry is relatively inefficient. (See the RICS/Reading University report for a comparison between the UK and USA industries).

One reason given for this is the clear division between the roles and responsibilities of the people involved. A linear approach to design, production and performance, in that order, is perpetuated, with those people involved in each of the three areas only brought in at a stage directly pertinent to them. The division of roles and responsibilities tends to create an arena for conflict rather than co-operation. If the computer is used as the focal point for all information, data, decision, recording, the generation of further information, etc. then it will be possible to have instant access to the project data. One computer and master program can serve all needs.

On this basis the following scenario is possible. A client produces a brief and asks the architect to act as project leader. Initial designs are produced and fed into the computer. From this it is possible to produce a feasibility study based on site conditions, cost, life cycle costs, energy demands, etc. If this is practical and within budget, then more detailed designs can be fed into the computer. Consultants can now be brought completely into the picture and they can feed in information, programs etc. to produce detailed drawings and calculations. As the design progresses, a fuller picture will emerge and the computer will analyse the new data to show the consequences of its utilisation in the design. At this stage it will be possible to bring in a builder to assess buildability. Any modifications can again by analysed by the computer. This will be the stage at which site work can commence.

If the consultant builder is the producer he will be familiar with the design and associated drawings. Again, using a data-base in the computer, a schedule of quantities can be produced. If the contract is awarded on a competitive tender basis, then quantities will be readily available. Overall medium and short term programmes can be produced and constantly updated: this information will be available to all connected with the project. The computer can produce three-dimensional graphics to aid the builder in the visualisation of the plans, etc., to aid construction. Costs can be checked and controlled, and interim payments can be easily formulated and agreed by comparing actual progress to programme costs, which are all on the computer. In effect, the representative of the project is the computer. It is able to hold all the relevant information and produce any data required. Using the computer, a high level of integration can be achieved between the design and production stages.

There are many drawbacks to this scenario. There will be problems relating to computer access. Who has it? How do you get information in and out? What about confidentiality? who takes responsibility for construction modifications punched in? Who pays for it? How many stations should there be? Where are the stations? Is it compatible with other computer systems? What happens if there is a computer failure? Who can have copies of printouts? Who keeps the program after contract completion? Who does the initial and subsequent programming? Whilst the prospects of using computers may be exciting and lead to greater efficiency, with decisions being based on quantifiable data, there will be those who will resent the apparent control which will pass to the computer. The openness of information flow might also be unacceptable. What is happening, however, is that designers who are already using computers for draughting and producing drawings are becoming more efficient than those using traditional skills. As clients grow in awareness they will inevitably demand a greater level of efficiency. It may be that, by harnessing the power of the computer, the designer, in conjunction with consultants and builder, will be able better to serve the client. The effective use of computers in design analysis will undoubtedly improve building technology.

SUMMARY

The growth, availability and use of computers can be of significant benefit to the design process. Quantitative data can be quickly calculated to give clear indications regarding the performance of materials, components and technology systems. The computer can aid design, produce quantities and schedules and monitor costs. In the future the computer may become the central information base for project data.

QUESTIONS

1. Will the use of computer-aided design produce a better designed construction detail?
2. Discuss the possibility of using a site-based terminal to draw information from the architect's master design program.
3. Who, and why, should control the master program if one is created for a project?
4. Describe a system of computer-aided design and a system of cost programming for a builder, the two of which could be combined. What advantages would accrue?

6. Constraints

Whilst there are many disadvantages in having professions in the construction industry, e.g. the maintenance of practice standards, there are also some drawbacks. These include their separation into individual bodies, making communication between them difficult. Design for example is separate from production, which in turn is separate from performance.

In any field there are a number of representative organisations such as trade associations and trade unions, but compared to other industries, the construction industry is highly fragmented. In the car industry design and production are contained within one organisation; there are some professional bodies but they do not impinge upon the process of making motor vehicles; farmers, to take another example, are represented by the National Farmers Union – it speaks with one voice. In construction the Group of Eight, consisting of representatives from employers, the professions and unions, attempts to create a dialogue with government. Individual organisations, who naturally put forward analyses based on their own perceptions and interests, further government by lobby. The Group of Eight is a reactive rather than a proactive body. Perhaps the industry ought to look within itself for improvements before trying to convince government on specific issues.

Such separateness makes the integration of design, production and performance very difficult. Until recently architects were barred from taking up directorships in building companies although they could be employees. Clients have tended to employ architects directly, having little opportunity to select them on a competitive basis. From the appointment of architect onwards the construction process tends to be a linear progression.

Technological decisions are seen as the responsibility of the architect who sets the framework within which the others have to work. The mould is formed early and other participants have to shape their contributions accordingly. With the merger between the RICS (Royal Institution of Chartered Surveyors) and the IQS (Institute of Quantity Surveyors) we have seen an integration between design functions and production functions, as far as the tasks of measurement of building work and cost control are concerned. Membership is not restricted by the nature of an employer's business role in the industry, only by an individual's ability to meet the membership criteria.

Attention should be concentrated on ensuring that all parties work closely together, at high levels of communication, in order to produce buildings to the required standards, giving value for money. The industry is piecemeal in its reaction to change: one body advocates, a second disputes, whilst a third carried out the work to inadequate standards. The client suffers as the industry perpetuates its image of giving poor service.

Why is it that change is relatively slow and does not seem to respond to demand? The main resistance probably comes from the reluctance of any organisation to cede any part of what it sees as its traditional role, or the representation of its members' interests. No organisation is willing to carry out radical change, let alone disband itself.

The focus must be on the creation of the building. If this is kept firmly in sight by all parties then ways and means ought to be forthcoming to achieve improvements without undue loss of status. It may be that existing boundaries will need to be crossed but that is a price that will be worth paying to keep the industry productive and of benefit to the nation.

Manufacturers play a prominent role in innovation. They have to market their goods on the basis of technical performance; quality and cost; availability and delivery; and literature/back up service. The latter is the means of communication to specifiers and buyers. This situation can impose a further constraint upon effective technology. Both designers and producers need to be convinced of the ability of a product to satisfy their own criteria. The need for clear, full and correct technical information was illustrated in the case history of the Summerland fire. It is only by thorough technical literature that designers and buyers will obtain some idea of the relevance of a product. Here is another communication gap that needs bridging, a constraint in itself. With control of the design and manufacture of materials and components in the hands of the industrialists, neither designer nor builder can exert much influence. This is another constraint upon design: the architect has, in effect, to take what is offered between alternatives. To design and produce afresh for each particular component is costly in time and money. We look at these problems later when considering component construction.

SUMMARY OF PART TWO

Firstly, the different approaches to design and its objectivces were considered, and perspectives of the participants in building design. The introduction of quantifiable and science-based techniques was explored and their contribution to an optimal solution was shown. A case history was given which illustrated some design procedures, their failure, and the subsequent consequences. The problems affecting the design production interface were highlighted. The use, or otherwise, of computers was investigated. Finally, some attempt was made to consider the constraints imposed on design.

Fig. 2.6 on p. 83 showed how the different techniques and approaches can be reconciled and integrated to produce one comprehensive approach to design.

Design is now not just the concern of the architect. With the increase in complexity of the technologies, and the resultant plethora of specialists, the design process becomes one of integration rather than of pure creation. The builder is contributing to decisions alongside the services engineers, the structural engineers and other specialists.

In some cases the builder provides the whole of the design/build function, whether it is under the package deal of a contract where the client takes a standard building, or where the builder acts as project controller, appoints the designers and manages the whole process. The scene is set at the design stage, which is accordingly the most important phase of the construction project and one that warrants further analysis and improvement. The builder needs to fully understand this part of the process, especially where he is not part of the initial decision-making.

QUESTIONS

1. How can the builder contribute to the design process?
2. Does the introduction of scientific aids to design necessarily impose extra costs?
3. Taking a particular construction detail, demonstrate how it might be improved, using the techniques of analysis described.
4. How far should the builder comment upon the appropriateness of a construction detail if it is suspected that it will not perform adequately?
5. To what extent is design a thinking process as compared with a commitment to meet certain procedural stages?
6. How far should design be integrated with production?
7. Should specialists be totally responsible for the design and construction of their particular contributions?

Part Three
PRODUCTION

1. Introduction

This Part, relating to the production of the building, is concerned with strategies and factors and their consequences for the technology of a building. There are two main themes running through the issues discussed – the design/production interface and the factors affecting production. Deeper consideration will be given to the manufacturing processes involved, not in technical detail, but as alternative strategies. This leads to the concept of component construction, where a developing trend can be seen for buildings to be supplied as a 'kits-of-parts'. This will affect the way that buildings are erected. The issues are set out in three chapters – on Manufacture, Components, and Assembly – which should be regarded as an entity.

The production of a building cannot be independent of the design nor of the technical/economic/social factors currently prevailing. The concern for the design/production relationship and its consequences can be encapsulated in the term 'buildability'. It has already been shown in the section on design that design decisions can directly affect the ease, or otherwise, of construction. In the question for buildability the fundamentals of good design, function and performance should not be overridden. A procedure that may produce an easy to build element could sacrifice, or lower the standard of a performance criterion. This is unacceptable and in such a case a relatively inefficient construction method would have to be adopted.

In theory it is possible to design and produce buildings that can be put together quickly and easily. Indeed such examples exist already, such as single storey steel-framed lightweight clad factory/warehouses. The main work is in the construction of the foundations. The frame and cladding is brought to site prefabricated, and assembled in place. This process consists basically of bolting together the units, whether frame or cladding. The design is such that shapes and lines are kept as simple as possible, bearing in mind aesthetics, and therefore standard sizes and profiles can be used, reducing the cutting and fitting operations. In addition, this type of construction is normally carried out by specialists, thereby ensuring a high level of quality control and avoiding delays caused by operatives 'learning' the method of construction.

There may be problems where the method employed, although

apparently simple, is innovative and therefore new to the operatives. If it is unfamiliar to site personnel, time and trouble may be expended in learning and adopting the process, thereby reducing any possible savings. In the case of a one-off structure, the whole of the estimated savings could be lost as there is not the opportunity to use the technological knowledge gained on subsequent buildings.

Gray has carried out some research into means by which productivity can be improved with respect to buildability (C. Gray; 'Buildability – the construction contribution'. CIOB Occasional Paper No. 29, 1983). His main finding is that the greatest benefits are to be found in simplifying the sequence of tasks, rather than simplifying the actual task carried out by the operative. He states that there is a fourfold return on improving the sequencing over that on improving the task itself.

It is therefore to the design that one should turn first to see if it can be amended to produce a simpler sequence of work packages. In their turn these can be examined with a view to improving the contributory tasks. To be effective the construction contribution must be incorporated at the very earliest stages of the design process, ideally within stages A and B of the RIBA plan of work. The problem lies in clearly identifying the work packages. A design is a composition of known categories of components and construction practices which can be joined together in a limited way to achieve an infinite variety of buildings having a wide range of uses. In the analysis of a design, to seek the work packages it is necessary to establish the particular components; their particular arrangements; their combination. It is at this level that improvements to gain in buildability can be most effective. The complexity of the combination of the tasks and their interrelationship need to be identified. Gray has put forward a model which can aid the appreciation of the complete network of interrelationships. This will give information upon which cost calculations can be based and their consequences extracted.

A work package needs to be at a level which is easily identifiable and can be differentiated from another. For example 'concrete work' is too broad as it does not distinguish between walls, floors, frame etc. 'Placing reinforcement in beam' is too narrow. The work package must also reflect the significant differences, so that such items as material, trade and plant are included in the overall description, which is presented in tabular form.

The next stage is the calculation of the duration of the work package. This is dependent upon the key resource in that package, such as speed of operation (or availability) of a piece of plant. Consequently the 'pace' work packages can be identified. These are the ones that are seen to control the overall rate of progress of the project as initially designed. They are the links in the continuity of resource utilisation. This structure of relationships between work packages can then be used to generate a model of the construction programme, which Gray calls NPS (Network Processing System). This is designed to operate at a minimum level of detail but gives an accurate forecast of the time needed to construct various parts. The work packages must be identified and the relationships between them given as relationships and not as fixed periods. In other words they must demonstrate how they depend upon each other to create the sequence structure, not solely on their individual and collective time periods, as demonstrasted by a critical path network. As Gray says

> By careful analysis of a design into its constituent work packages it is possible to test the construction implications of alternate design strategies by comparing the total implications of the alternative work packages and their combinations. This, however, requires an ability to select the work packages and calculate their relative importance as the design is forming, probably even before it is committed to paper, because as soon as a design starts to be drawn it is physically and psychologically extremely difficult to amend and the commitment to cost is made.

An independent study carried out by NEDO (*Faster Building for Industry*, 1983) tends to support Gray's concept that improvements in buildability can be achieved by considering the relationships between work packages, rather than looking at tasks in detail at different skill levels.

The report describes a study of the overall speed of erection of industrial buildings, which compares differing contractual structures, such as design/build, project management, traditional main contractor/sub contractor and so on. The main conclusion was that, on average, project management arrangements produced faster building at no extra cost to quality and performance. Those contracts which were let under a main contractor and specialist sub-contractors tended to be the slowest, with a high number of these extending beyond the stipulated contract time. The problem can be seen as an inability to achieve optimum interrelationships between the work packages as encapsulated by each specialist sub-contractor. Each sub-contractor was able to meet the requirements of his overall time on the site. What was not effectively achieved was the optimum overlap between the work packages, which would have contributed to an overall reduction of the contract period. Gray says that it is the interrelationship between the work packages that is significant, not the time period in itself.

Whether it was a problem in design which created the inability of the contractor to co-ordinate and interrelate the packages, or just poor management, requires further investigation. In the latter situation it may be that incompetent production sequences and methods have been adopted by the contractor which lead to an increase in contract time. Through proper project management two advantages could arise:

1. As Gray demands, a contribution could be made at design stage which will ensure a near optimum interrelationship of work packages.
2. The level of expertise in project management is such that the production processes could be well through and vigorously controlled.

Whatever the reason, the major point being made here is that the technology of the production process affects buildability.

The Construction Industry Research and Information Association (CIRIA) in *Buildability: An Assessment*, 1983, has attempted to give some guidance on where aspects of construction could be improved. Fig. 3.1 lists the seven general principles. Most of these are directed towards the designer but necessitate the involvement of the builder.

1 carry out thorough investigation and design
2 plan for essential site production requirements
3 plan for practical sequence of building operations and early enclosure
4 plan for simplicity of assembly and logical trade sequences
5 detail for maximum repetition and standardisation
6 detail for achievable tolerances
7 specify robust and suitable materials

3.1 General principles for good buildability (from Buildability: an assessment, *CIRIA Special Publication, 1983)*

1. *Carry out thorough investigation and design.*
A four storey block of flats was to be built on the site of a demolished cinema. On breaking up the ground floor slab a large basement housing an obsolete heating boiler was found. A more thorough investigation would have revealed this earlier.

2. *Plan for essential site production requirements.*
An office building was designed with large precast concrete panels as an external cladding. It was a restricted site with access to one side only. It was found impossible to lift the panels to the far side

of the building with the plant then currently available. The panels had to be remade in smaller sections.

3. *Plan for a practical sequence of building operations and early enclosure.*

A building had an in situ reinforced concrete staircase meeting with a precast floor, both supported on brick walls. Half landings were supported by brick walls so that concreting was dependent upon wall erection before and then after up to next floor level.

4. *Plan for simplicity of assembly and logical trade sequences.*

The use of plasterboard dry lining was not compatible with internal room dimensions. Much cutting, fitting and subsequent waste was incurred.

5. *Detail for maximum repetition and standardisation.*

A variety of window opening widths meant that a large number of different sized lintels had to be manufactured.

6. *Detail for achievable tolerance.*

An in situ concrete framed building had in situ floors of precast concrete panels with applied tile finish. Metal window frames were fixed to the in situ frame and resting on the precast panels. The tolerances for each of these were incompatible in that in their worst condition a gap tapering from 35mm to 3mm was possible between window frame and internal cill.

7. *Specify robust and suitable materials.*

Wood wool slabs specified as permanent formwork for concrete did not perform satisfactorily as they absorbed vibrations for compacting the concrete.

The CIRIA study summarised the assessment with two major points:

1. when good buildability has been adequately defined and developed, it leads to major benefits for clients, designers and builders;
2. the achievement of good buildability depends upon both designers and builders being able to see the whole building process through each others' eyes.

It is to be hoped that these conclusions will be noted by all parties to the construction process and incorporated fully into practice.

Production processes, whether on site or off site, are dependent upon many external factors in addition to those directly related to the building's design. Here it is necessary to bring into the equation the contextual framework factors, in particular economic, social and technical factors.

In general terms construction activity is bounded within prevailing social and economic conditions. For example, there is a growing proliferation of specialist sub-contractors operating within the industry. (It is not intended to discuss here the pros and cons of the use of sub-contractors as opposed to directly employed labour. What must be realised though, is that in any consideration of the methods of production the status of the operatives must be taken into account). This is grounded in economic reasons and social expectation. Government statistics have mapped a decline in the number of operatives employed by construction companies (partly due to the recession of the late 1970s and 1980s) but this is counterbalanced by a massive increase in the numbers registered as self-employed or in firms of two or three. There is a social status in being self-employed, in addition to any possible economic benefits, which may be a contributory factor in the swing towards self-employment. This increase in self-employment is supplementary to the growth of specialist firms. Specialist firms are increasing because of the growing technical complexity of buildings, not purely as a result of social and economic factors. As manufactured elements, components and materials take up an increasing part in the overall construction, then it follows that there will be a corresponding increase in fixing specialists. These may be under the direct control of the manufacturers; fix under licence; supply and fix; or be independent. Whatever relationship pertains, the construction processes on the site will be affected in some way. If the manufacturer supplies and fixes, then the builder takes on a co-ordinating and supervisory role in checking the technology.

The builder does have any direct control over the methodology of the installation process. What is expected is that the work should be executed within the overall periods set by the builder and to the quality standards laid down in the specification. The prime responsibility rests with the manufacturer/supplier/fixer. The independent fixer has a responsibility only for correct installation. The builder has to satisfy himself that the goods are supplied to the right specification and are properly unloaded and stored. He will have greater control over the site operations and in the detail of methodology and timing of the operations. With directly employed fixers the degree of control can be even greater.

National factors can influence production processes, as seen in the 1960s with the use of industrialised high-rise building systems. The majority of panels were manufactured off site and erected on site by firms under the control of the suppliers who were responsible for quality control. These systems were introduced to satisfy a number of economic and social requirements:

1. to provide mass housing after clearing inner city slum areas;
2. to upgrade existing standards of accommodation;
3. to keep to a minimum the amount of land per dwelling;

4. to build quickly and to the least cost;
5. to decrease the amount of on site work which can be uncomfortable, dangerous and subject to the vagaries of the weather.

Whether or not these objectives were met is debatable; the weight of argument is firmly of the opinion that they were not.

The factors which influence production strategies within the orbit of the builder are listed below, not in any order of priority.

1. *To build quickly.*
The thrust of the strategy was to ensure that a large number of dwelling units was erected as quickly as possible. Construction methods were geared to the reduction of construction time on site. By placing some of the work off site a decrease in the overall time was achieved.

2. *Effective use of resources.*
To build the targeted number of dwellings required a substantial increase in the work force. The number of skilled operatives available was not enough to meet demand. To train a person to a competent level of skill took four years, too long to make an impact on the construction programme. Therefore, an alternative method of production was sought, viz. factory production. This could be controlled by relatively few skilled people and with automated machine production lines. The on site fixing skills were reduced to a minimum, although a high level of care was required. The number and range of on site plant and equipment did not increase dramatically. If building a tower block with in situ materials, tower cranes, hoists and scaffolding were used; the same was the case for prefabricated systems.

In the British building industry it is traditional to use concrete. Most design knowledge, practical experience and development was centred on concrete rather than steel. Concrete was thus the best raw material to use for prefabricated systems.

3. *Scale.*
Government policy during this time was to set targets of a number of new dwellings per annum. Builders and suppliers could see a future market of tens of thousands of units (national targets were in the order of 300,000 dwellings per year). It would be worthwhile investing capital, labour and time in the provision of factories for the production of concrete building systems. Economies of scale could be anticipated.

4. *Economics.*
Client and builder alike want a building which satisfies their needs according to economic criteria. The client requires the least cost, commensurate with brief. The builder requires the maximum

profit in meeting that brief. There is little doubt that a major contribution to the promotion of industrial building systems was the profit motive. Unfortunately, the number of system dwellings had to be drastically cut after the Ronan Point disaster in 1968. This also coincided with a general reduction in the overall target figures. Large factories able to produce many units were surplus to requirements and many companies were liquidated or turned their attention to other forms of construction. The long-term profits did not materialise, although they were there in the short term.

As regards the economies of scale to the client the system building units did not become cheaper than traditional methods until 1970/71. By then the number of units being built was in rapid decline.

SUMMARY

The concept of buildability as the interface between design and production aspects, and its promotion, can be of benefit to client, architect and builder alike.

In looking at the rise and fall of 1960s system building the social and economic factors affecting production strategies were highlighted.

QUESTIONS

1. From your experience identify examples under the seven headings given, where buildability could have been improved.
2. Discuss how the builder can make a major contribution to the achievement of buildability.
3. To what extent were the system building suppliers responsible for their own eventual fall from grace?

2. Direct Factors Affecting Production

The direct factors are centred on the building itself and its location. One can identify, in general terms, these factors, but cannot forecast their influence on construction as this will not be constant. They will vary as a result of the social and economic conditions mentioned previously and according to the perceptions and expectations of the initiators of the building.

THE STRUCTURE OF THE BUILDING

The nature of the structure must be the dominant factor in determining the production of the building: that is, whether it is supported by a frame, load-bearing walls, core, suspension etc. We will take two examples, those of frames and load-bearing walls. The method of production for frames is dependent on joining together columns and beams. A complete frame can be erected, leaving the floors and walls to later. Load-bearing walls carry themselves, so are limited in height. Lateral restraint is required for permanent stability and is difficult to make provision for the placing of the floors after the next storey section of the wall. Special nibs or notches are required to give support, which increases the complexity of construction. Therefore, walls and floors are built together in storey height stages.

MATERIALS

The nature of the materials used in building elements will influence the methods of production. Again, taking the examples of frames and load-bearing walls, one can further differentiate within the individual structural system. If the frame is in steel the production is based on a crane lifting the units into position with the steel erectors fixing and checking. If the frame is in in situ concrete then a completely different set of technologies are employed. Formwork, steelfixing, concrete placing (transporting and mixing), together with access platforms are required. Although both materials are finally in frames the production processes are very different.

Similarly in the consideration of materials for load-bearing wall structures, one may compare bricks with precast concrete panels. A brick wall is built up item by item, using mortar. Special skills are required, also access platforms and means of protection from the elements. Precast concrete panels need to be lifted into position and fixed. Each process depends upon a dissimilar resource, brickwork on specialist placing skills, concrete panels on lifting machinery.

SPEED

A building may need to be erected at a faster rate than normal. This may be due to the need to fulfil an important user function, or to cause the least disruption. If there is a need to, as the Americans put it, 'fast track' then the overall production process will have to cope. Operations will have to run concurrently rather than consecutively. Special plant and/or machinery, and temporary supports, may also be required. Some components or elements may be assembled off site.

SITE CONDITIONS

Here there are two considerations:

(a) the geographical/geological constraints of the site;
(b) those constraints imposed by the position of the site with relation to the environment.

The first aspect is centred on the nature of the site: the soil type and conditions; the topography. Poor soils will demand complex and extensive foundations which will require complicated methods of production. Sloping sites will pose problems which will demand specific production solutions.

The general environment of the site will affect the processes of production. If in a crowded urban centre restrictions on unloading facilities, constricted approach roads and so on will influence the size, weight, quantity and type of materials and components. Serious consideration will need to be given to these aspects in the design of the building.

SIZE OF BUILDING

If the building is large, either horizontally or vertically, it may lend itself to prefabrication of its major elements. Some degree of standardisation could be desirable. A small one-off building may best be built using traditional means as the cost of large plant and equipment could be too much of a capital outlay.

QUALITY

The standard of finish or general quality of workmanship can affect the production process. Obviously the highest levels of quality must be sought at all times, but these must be appropriate to the specification and estimated prices. Value for money should be sought by both client and builder. Where there is an expectation for high quality it does bring into focus the need for greater care in handling, fixing and protecting the relevant parts. Production methods will need to take full account of the need for extra care and the possibility of a general reduction in the rate of working. Further operations may be required to ensure that the quality levels are reached and/or that the finished product is protected.

THE 'WHOLE BUILDING'

A building which in its entirety may be innovatory, novel or dissimilar to others will have to be considered as unique. Most buildings can be said to be unique, but the majority are modelled on conventional designs, shapes, materials etc. In the case of the unusual, production technology will need to be directed totally to meet the criteria posed by the design. The opposite may also occur, that a system of production will only be possible if the design is modified. In other words the production methods will determine the shape, size etc. of the building.

An example of a 'whole' building concept influencing the production, and the production in its turn influencing the final outcome of the building, is the Sydney Opera House. The initial concept put forward by the architect, although accepted and desired by the clients, was found to be impossible to build. Major modifications had to be made to meet the demands of the production process.

DESIGN DISCIPLINE

There are many different schools, ideas, disciplines, methods, styles in design, resulting in a variety of approaches to solving design problems, as shown in Part Two. The type of approach may distinctly affect the production of a building. For example, a design based on performance criteria will place great responsibility on the builder to select and use suitable components, etc. The performance criteria may not be solely related to the 'in use' aspects but could also lay down expectations for the placing and fixing.

AESTHETICS

Aesthetic considerations could be applicable to the whole building or just to sections or parts. The inclusion of curved soldier arches over window openings will demand careful consideration of the techniques for achieving these. The requirement, of say, exposed aggregate finishes to concrete will demand a variation in the normal procedures.

COST

Finally, one must never underestimate the influence of cost upon the process of production. Many would say that this is the main criterion upon which all production decisions are made. In the consideration of alternatives, where each meets the criteria, then it is cost which determines the course of action. But before costs are calculated the process must be determined, which is governed by some if not all of the foregoing factors.

SUMMARY

The above ten factors have defined the envelope within which production decisions are formulated and realised. They must be applied afresh to each particular building. Every construction enterprise should be appraised using these factors to ensure that optimum production methods are adopted to benefit both client and builder.

QUESTIONS

1. Which of the above factors are within the control of the builder?
2. Compare two internal wall finishes and show how they can determine production methods.
3. Taking a simple single-storey factory built of load-bearing brick walls, rank order the above factors for their effect on production methods. Compare to a prefabricated steel frame with lightweight cladding, again using a rank order for the factors. What are the reasons for the differences?

3. Manufacture

Where once many items were made on site, now they are manufactured in factories and workshops. Common examples are lintels and door frames. The reasons for this change stem from both technology and economics.

TECHNOLOGY

As the nature of materials and the design of buildings are more and more founded in scientific analysis and knowledge, there is a corresponding realisation of the need for greater quality control and higher standards. On-site working by its very nature is not conducive to the reproduction of the highest standards possible. Most activities are carried out in the open air, with little control over the climatic conditions. Space for making and subsequent fixing is usually at a premium. Fabrication on site requires the additional resources of labour to carry out that part of the process. The overall length of the task has to take into consideration the fabrication period, which can extend the construction time. The type of machinery and plant available on site cannot match factory installations in meeting high levels of tolerance.

ECONOMICS

Following hard upon these factors are the cost implications relating to the decision to manufacture on or off site. If there is a strict comparison made between a unit made on site to one made off site, generally the former is cheaper. This is mainly due to the fact that on-site overhead costs in plant, machinery, premises and administration are lower than those in capital intensive factories. The advantage in factory production is that economics of scale can be achieved (assuming there is a large enough market) and that time can be saved on site. In other words the overall project period can be reduced, thereby reducing construction overheads.

Although the decision to either make on site or buy in is primarily taken by the designer and/or builder the alternatives are becoming less clearcut. Component manufacturers are developing and marketing a great variety of new items. In many cases they do not just replace a site-based activity but make it far easier. Take again the case of lintels. With the development of lightweight steel sections concrete units are less popular. It is now possible to have the cavity tray incorporated into the lintel as well as a flange for supporting external brickwork. The lintel is easily lifted into place and can span wide openings. Another example is the fixing of door and window frames to external walls in brick-block cavity walls. Extruded plastic sections are nailed to the back of the frames to provide a vertical damproof course and a seal to the open end of the cavity, a groove for the placement of wall ties enabling the structural fixing to the walls. This demonstrates a dramatic improvement over the traditional method of placing elm pads in the brick courses, creating brick returns to seal the cavity and placing a vertical roll dpc (in addition the frame had to be nailed to the elm pads).

With the advent of these new and markedly better items and components the builder and designer would be at a disadvantage if they did not use them. These particular examples can show considerable savings on site costs. Not only that, but they exhibit a higher standard of quality in precision, durability, structural integrity and in solving a number of technological problems within one component.

The advantages of off-site production have been enhanced by the use of man-made materials. Traditionally, buildings have been constructed mainly by using natural materials commonly available. Although some work has to be done in rendering them suitable for construction, the technology for these processes is relatively simple. Timber is easily converted into suitable sections, lengths and finishes, bricks are fired from dug clays, etc. Many buildings still use a preponderance of these materials, but there is an increase in the use of materials which require factory production techniques. Two which immediately spring to mind are plastics and steel. Plastics are making their impact in three areas of use: in pipework, small components (e.g. door furniture) and finishes. In each case it would be impossible to shape, let alone make the plastic on site. With regard to steel, advances have been made in the development of sections other than universal beams. Now available are box and round hollow sections. As steel sheet can be easily shaped in the factory a variety of profiles and shapes can be produced to suit most building configurations. In the case of both these materials the manufacturing processes can be controlled to ensure the highest standards of quality and tolerance are met.

Not only can the individual materials be successfully produced but they can be combined to give advantages over and above their previous qualities. The use of steel sheet with a plastic coated membrane as a finish, including an insulation material, is an example of a use as external cladding. It is lightweight, easy to fix, self-finished and durable. It is available in a variety of profiles and these can be bent to give rounded corners. The combination gives a high quality component which can exceed the performance factors of traditional materials. This leads to another factor which plays a large part in the trend to factory production of building components, namely performance.

PERFORMANCE

One of the goals of most human enterprises in their quest for improvement is the ability to predict performance. Predictions for performance of materials are more accurate if they have been produced under controlled conditions. Testing and modifying at the place of manufacture is easier than at the construction site. Many material and component manufacturers have their own research departments whose main function is to test existing materials and develop new ones. Most innovations arise from work carried out by manufacturers in an attempt to solve a technological problem. (See Bowley, *The British Building Industry*, 1966.)

A major problem in the assessment or prediction of the performance of a material or component, is the evaluation of performance with time. With a component or part of an engine or machine a test bed approach can be used. Empirical tests are carried out, to failure if necessary, to build up knowledge of performance characteristics under working conditions. This approach is extremely difficult to emulate with construction components. Whereas the test bed simulation can be speeded up to condense the expected working life, this is not possible when assessing the weathering characteristics of a new cladding panel. Even if we could simulate the changing seasons, temperature changes, rain and snow fall, solar radiation etc. these could vary considerably according to the position of the panel. Its exposure is affected by: height (above sea level and of the building itself); adjoining buildings; overall surface area; type of joint; type and proximity of other materials.

The tests and monitoring procedures of the Agrément Board are geared to evaluating the component in use over a long period of time. Unfortunately, not many items have been offered to the Board for assessment, just a few hundred per annum in the mid-1980s. It is not mandatory for a component to have a certificate issued by the Board in order for it to be used in any building; as long as it complies with any relevant British Standard or Code of Practice then it is acceptable to most specifiers. British

Standards lay down tasks and procedures for the fitness of a material or component in its own integrity. Tests are confined to the quality and performance of the item as manufactured. An analogy can be made with the human body. Tests can be applied which show that the body is fit, healthy and strong and capable of carrying out physical activities. The tests will not however indicate whether or not that body can climb a mountain and perform under conditions of stress and discomfort. That can only be ascertained by monitoring during the actual activity. The assessment under the Agrément Board's auspices is similar to the monitoring during a climb. This branch of performance assessment is relatively new and still in its infancy. But the results of these long term in-place tests are now being understood and being given significant recognition. Their continuation and expansion to include all building elements, materials and components should be encouraged in order to promote the highest standards, as the 1985 Building Regulations have stated.

What are the possible benefits to performance if production is placed in the factory? In theory, the main benefit should be an overall raising of standards. Whether or not these are achieved depends primarily upon the quality control procedures exercised within a factory. In addition the outcome of the process should be predictable. For example, the mixing and casting of concrete should be easier and more reliable under factory conditions. Sophisticated and sensitive machinery can be used; transportation can be rationalised; temperature can be controlled; clean and accurate moulds can be formed; placing can be carried out in ideal conditions; curing can be strictly monitored.

The criteria for production ought to be considered from the point of view of the manufacturer. As already stated, it is not the builder who makes the decision whether or not to set up a manufacturing process for the production of components. Even industrialised building systems were developed by specialist companies and promoted and sold in partnership with builders. Bison Wall Frame was erected under the main contract of different builders; the Larssen Neilson system in partnership with one builder. We must therefore look upon the criteria for manufacturing as being different from those for the production of buildings.

MATERIALS

The nature of the materials used for a product determines the manufacturing process. It is what is done to that material which transforms it into a usable product. Its availability should be reliable and to the required quality. The manner of its handling to and around the factory must be taken into account. If used in conjunction with other materials compatibility should be ensured.

PREMISES

Are special buildings required for the manufacturing processes? Can existing ones be easily adapted? Will internal layouts restrict production flow? The location of the building can be important if the processes produce a possible pollutant; it needs to be away from urban areas. Also location can be critical if the products require good road or rail systems for transportation. The administrative function might benefit (and vice-versa for production) if situated close to production. Quality control could be improved if adjacent to sales, etc.

PROCESS

Between the material and the product is the process. The process is dependent on:

(a) number of units
(b) rate required
(c) complexity of manufacturing tasks
(d) shape, size and mass of units.

Some can best be made by machinery others by hand. Is the skilled labour available or can it be trained? Are there any problems in handling and storing the basic materials and the finished product? The quality control procedures should be able to meet the design criteria without being too costly.

PRODUCT

Before any product is manufactured the possible market must be ascertained. Many companies carry out market research to discover if the product will sell, and if so, its potential. This can depend solely on cost, especially if it is innovatory or has to compete with similar items. If the product is innovatory it must be seen to be useful; in other words to solve a design problem. It is important to know who will buy it. If the trend is towards contractual arrangements like project management by builders it may be that they will be the prime customers. Under the traditional linear form of construction organisation the architect may be the main specifier/buyer. Even if not actually buying the product he will make the decision as to its use. Building users are now taking a much closer interest in the choice of products. They are concerned about the long-term performance of the building and its parts. Assurances regarding durability and adequate functioning are commonly sought, with an increasing responsibility vested in the manufacturers to rectify any faults at their own cost.

Notwithstanding the last remarks the designer and/or builder should satisfy themselves of the performance characteristics and suitability of the product for the particular building. The product must work in place. If the product is one that is of major importance to the building, whether structurally, aesthetically or functionally, those responsible for its procurement should ensure that the production process is able to meet their requirements. For the architect these could relate to colour, quality, finish etc. and for the builder, rate of production, delivery and methods of handling and tolerances. A visit to the factory would give some reassurance (or otherwise) in this respect.

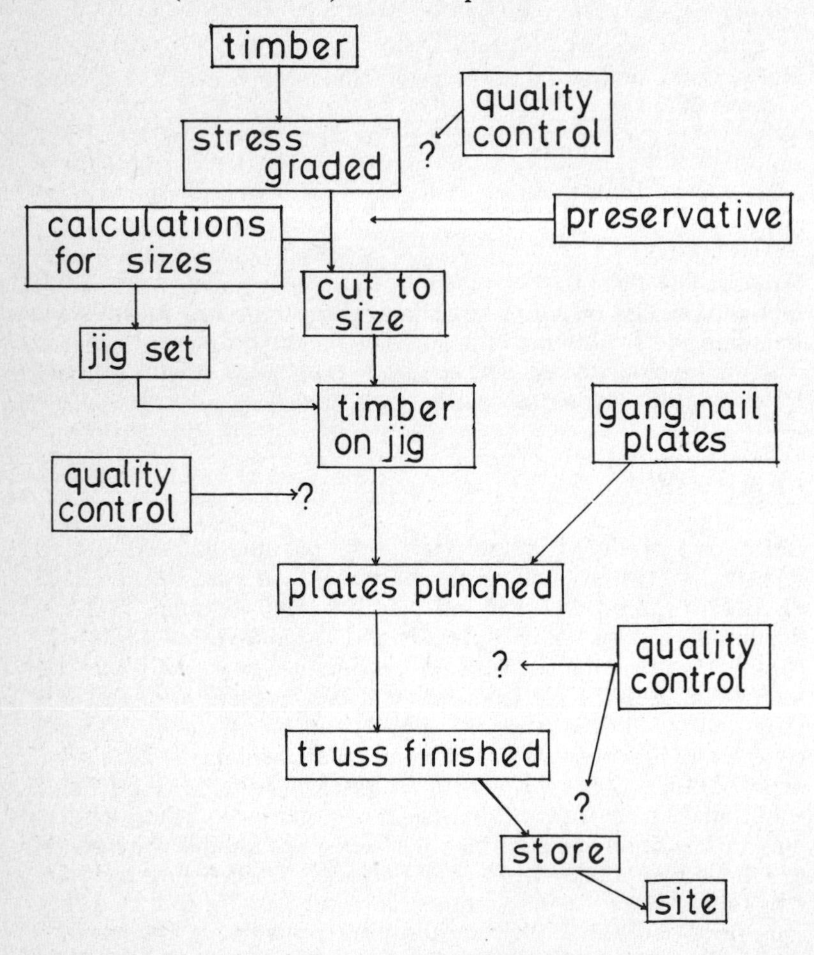

3.2 Manufacturing process: trussed rafters

A common product which is typical of the movement towards manufacturing products in off-site factories is timber roof trusses. Whilst they are successful, functional, easy to fix and cost-effective they have also suffered from faults and failures.

They were first introduced, in quantity, in the early 1960s to coincide with the boom in house building. Their obvious appeal was that the amount of timber required for a pitched roof carcass was reduced significantly; rafters, struts and joists were in one unit; ridge boards were not required; two operatives could lift and place them relatively easily onto a two story house; a roof carcass could be erected in a fraction of the time of a traditional rafter, joist, ridgeboard and purlin roof. No cutting was required on site.

Monitoring and advising the erection procedures for trussed rafters are the BSI Code of Practice; the Trussed Plate Association; and the Timber Research and Development Association. A survey carried out in 1978 (Baldwin and Ransom, BRE Current Paper 83/78, 'The integrity of trussed rafter roofs') showed that in that year there were 700 cases of faults and failures. The great majority of these failures involved movements due to instability of the roof structure, the basic cause being inadequate bracing of the assembled roof, rather than the failure of individual trusses. Other problems encountered included under-design, unauthorised alteration and plate corrosion. A follow-up study (Mayo, Rodwell, Morgan, BRE Current paper 5/83, 'Trussed rafter roofs') concentrated on their manufacture, site use and performance in service, and the significance of the copper-chrome-arsenic preservation treatment in the incidence of corrosion of galvanised metal fasteners. Consideration will now be given to its findings in relation to the manufacturing process.

SPECIFICATION AND DESIGN

A problem raised by the fabricators was their difficulty in obtaining sufficient information from the customer to enable the design of the trussed rafters to be carried out exactly to the customer's requirements. In the absence of detail the fabricator made assumptions, such as position of hatch and chimney openings, trussed rafter supports, position and capacity of water tank, types of tiles used, gable details etc. But it is doubtful in many cases if the customer really understands the significance of these assumptions and his agreement to them serves no purpose other than safeguarding the fabricator, if subsequently the trussed rafters prove unsuitable.

Fabricators were reluctant to produce designs for the complete roof. They concentrated on designing individual units. This arose from the lack of information about bracing the roofs, other than standard domestic. Design details for the standard fink, fan and mono-pitch trussed rafters were normally based on the system owners design manuals. These gave relevant timber sizes with regard to span, pitch and stress grade of timber, together with cutting schedules, plate sizes and plate locations on the members. The design of up to 40% of 'specials' made was based on CP112:Part 3:1973 and carried out by computer. As the report stated:

> It was evident from the lack of visual grade markings that little formal visual grading was carried out, although most of the fabricators visited employed qualified visual graders. About 48% of trussed rafters delivered to site did not exhibit any indication of stress grading, either directly on the timber or from drawings, specifications of fabricators' delivery notes.

PRESERVATIVE TREATMENT

More than 70% of local authority sites used trussed rafters treated with preservatives, compared to less than 40% of private developments. It was clear from visits to trussed rafter fabricators that little effort was made to dry preservative treated timber or even allow adequate curing time before fabrication. Several cases were observed of recently fabricated trussed rafters dripping with preservative. Not only is this contrary to the advice given by the preservative manufacturers but it completely ignores the recommendations in CP 112:Part 3:1973 on both preservative treatment and the moisture content at the time of fabrication.

QUALITY CONTROL

Quality control was virtually non-existent. When checks were made they were random and on an informal basis. Timber was generally stored outside before and after fabrication, with no assessment of its moisture content, which should be less than 22% to comply with CP112:Part 3:1973. No checks were made after fabrication on plate position, teeth embediment or abutment of the members at the joints. It was found that despite this apparent non-compliance to the Code of Practice the quality was 'reasonable'. Only one of the twelve fabricators visited used a formal system of quality control: it was a member of the TRADA quality assurance scheme.

POINTS TO PONDER

It has been noted that there is an increase in the use of prefabricated components and units throughout the building industry. Unfortunately, those investigations that have been carried out show that the manufacturing processes may not be producing the highest quality goods. An assumed advantage of factory production is greater control of the process to ensure good quality: in the case of the trussed rafters this was found not to be happening. A further report published by the BRE (K. Fletcher, 1983) showed that an alarming number of goods did not comply with the standards laid down. That the fault does not lie entirely with the manufacturers is pointed out by Ritter (J. Ritter, 'The role of quality and its assessment in design'), in *Quality and Profit in Building Design*, ed. Brandon and Powell, 1984). Specifiers and users need to pool their experience to form clear views on the appropriate quality for the product which they will buy. As seen in the case of the trussed rafters the specifiers (customers) did not supply full information to the fabricators relating to the use of the trusses. It is incumbent upon the customers to make clear their requirements, insist they are met and ensure that they are fixed properly on site.

It is clear that the trend to prefabricate off site will not falter and that many more products will be produced to supplant traditional materials and techniques. The building technologist must satisfy himself that goods are to the specified quality and fit for the purpose.

QUESTIONS

1. Give reasons for the increase in the range and variety of building components.
2. How has the DIY market influenced the range of building products?
3. What are the main criteria for setting up a manufacturing process for building products?
4. Can one rely upon manufacturers to produce goods to the right quality?
5. Compare the criteria for the manufacture of a building product to those of the builder in the construction of a building.

4. Components

A. Bender (*A Crack in the Rear View Mirror*, 1973) has shown that construction techniques in the US are based increasingly upon the concept of contribution with components. A component can be identified as 'an item or unit which is brought to site in either a finished or part finished state ready for placement in position'. Examples range from door sets, to trussed rafters, to heat pumps, to precast floor beams, to complete bathrooms with fittings and pipework, to half houses. Excluded would be bricks, random lengths of timber, blocks, in situ concrete. By implication, components will tend to be manufactured off site by specialists, not by the builder. They will be available in most cases from a variety of suppliers and bought according to specification and price. The responsibility for their initial quality will be the manufacturer's. The builder will be responsible for their effective placement and integration with other components. The use of components as compared to shaping basic materials to fit the building has major implications for the production process. Its influence commences at design stage and follows through to performance in use. It becomes a form of technology in its own right. It creates procedures, practices, tests and controls to cope with the nature of the products. We will consider now some of the implications of adopting a component approach to construction.

MANUFACTURING CONTROL

Some of the issues relating to the manufacture of goods for building structures have already been reviewed. The main control is in the hands of the manufacturer, although as customers, building technologists can make a strong input to ensure fitness for purpose. Nevertheless, the product is finally dependent upon the manufacturing process.

The initiative to make and market a product is generally taken by a manufacturer. The customer (designer or builder) is then a consumer and only has control over fitness for purpose. The Grovers have argued (R.J. and C.S. Grover, 'An appraisal of warranties as a product support policy for the construction firm', in Brandon and Powell (eds.), *Quality and Profit in Building*

Design, 1984) that by using warranties on products (in components and the whole building) quality may be better appraised. If this is extended to buildings then it must include the components that make up the whole building. In that case suppliers would themselves have to give warranties. It is obvious that little consideration has been given to this idea, as few warranties are given over and above a minimum guarantee. As mentioned before, comparatively few Agrément Certificates, which are an indication of a product's integrity although not in themselves warranties, are given annually. A warranty is a protection to the customer in the event of failure. Depending upon the product and its degree of possible failure, the manufacturer repairs or replaces the defective component, up to a stated time in use. The standards now operating do not outline a course of action in the event of failure.

RANGE OF ALTERNATIVES

As more and more building items are produced as components there is the likelihood that more than one manufacturer will market a product. Therefore, a range of alternatives could exist, each having slightly different characteristics in specification and performance. In order to make effective decisions the buyer should know the range available, have full details of their respective specifications, and be able to compare them on like scales.

A study undertaken by the Institute of Advanced Architectural Studies at York University has shown that the majority of architects rely upon previous experience in solving problems on new projects. They only refer to technical literature when no clear solution presents itself. (D. Wise, 'Informing design decisions', in Brandon and Powell (eds.), *Quality and Profit in Building Design*, 1984). Additionally, it was found that there was a need to increase the designer's awareness of the need to consult written information as well as of the information itself. As Wise says, 'The rate of change of product and technical information with new design factors and constraints was rapid and designers could quickly be out of date'.

There is no reason to suppose that builders consult technical literature any more than architects when deciding upon the products to buy. With the traditional form of construction where basic materials are delivered to site and built into position the decision is primarily in the hands of the designer or builder. They are made to suit exactly. In the case of a product it may be necessary to use that which comes nearest to the desired solution. The decision making process becomes a little more complex when having to decide between similar products. In the final analysis the decisions usually depends upon price. A further aspect relating to range is that it becomes a two-edged sword. On the one hand there is generally an opportunity to choose from a selection of possible products (which can be seen as a benefit to the customer), but on the other there is a tendency to standardise the range of sizes. It is this latter aspect which is of concern to the architect. It leaves little scope for creating space, shape, appearance, scale etc. if all that is available, say, is one size of window frame.

To minimise production costs it is in the interest of the manufacturer to rationalise their range of sizes. To compete in the same market with a different type or quality of product another manufacturer is likely to adopt the same basic sizes. Some architects are against this trend as they say it would lead to uniformity of building appearance. In other words, their scope for individual design would be limited, whereas the builder would welcome the standardisation of sizes. If the same sizes were used on each project the operatives and supervisors would know exactly what to expect. Setting out etc. would be simplified and a higher level of buildability would ensure. Familiarity with the component range sizes would speed the construction process.

EFFECT ON SEQUENCING

The use of components will have an effect on the total process of planning and programming the building project. This effect will also be dependent upon other factors, the main one being contractual arrangement. Contractual procedures and their effect on the component concept of production will now be considered.

1. Traditional competitive tender

The common situation here is that the client appoints an architect who prepares the specification and working drawings. Competitive tenders are received and a builder appointed to carry out the work according to the drawings and, perhaps, a bill of quantities. Generally, the architect has specified the majority of components (maybe even nominated suppliers) and the builder's responsibility is to order and buy them. At this stage in the process it will be difficult to alter or reselect an alternative. The builder will have to ensure that the suppy and fixing of the components suits his programme. Orders will need to be placed quickly as there is the probability of long lead times before any possible supply to site. The builder will not have had an opportunity to comment on the effectiveness or buildability of the component. There is little room for involvement or improvement; where the builder is aware of an alternative which is better suited he cannot easily introduce it.

2. Builder as project manager

Here the builder is brought in at a much earlier stage in the design process, with a consequently greater impact upon the choice of components. Generally, the project management team works closely with the architect. The designer has overall design responsibility and prepares the drawings which give clear intentions. Specialists, consultants, suppliers and manufacturer are invited to submit their solutions. These are evaluated on the basis for fitness for purpose, performance, meeting demands of the construction methods, ability to meet time constraints and of course, costs. The component is seen in the context of the whole building design, construction and performance phases. It is more likely that the component will be integrated into the production process with greater precision.

As knowledge of the component is fully developed at the design stage it can be incorporated into the whole scheme. Its relationship with other components and aspects of the building can be investigated to give firmer assurance of its suitability.

Project managers have been known to make the final decision regarding the choice of component or construction method. The architect has produced a brief and performance specification, and as long as the component meets those requirements, whatever its source, it will be accepted.

The point at which a component is selected will have some consequences for the overall sequencing of the project, as has been illustrated above. The earlier it is selected, with a full recognition of its impact on construction method, the better 'fit' it is likely to have.

When compared with traditional methods of construction a distinct difference is seen in the degree of knowledge required of any detail. If it is built in situ with basic materials these can be gathered on site as and when necessary; there is a greater flexibility in their procurement. A component needs to be fully specified very early on and its availability time incorporated into the programme. If the lead time for its manufacture and supply is long, it may put back the site activity. There have been instances where a contractor has given a programmed time for the job which is less than that subsequently found to be necessary for one of the components' manufacture and supply. If production of the construction detail could be carried out directly by the builder then it is possible that the original programme could be met. When production of a component is realised off site it must be recognised that it could have an adverse effect on the sequencing and programming of the job. Certainly, more attention needs to be given to the availability of components as compared to the availability of basic materials.

RELATIONSHIPS

Here we are looking at the relationships between the main parties to the building process. When extensive use is made of components in construction an additional party is brought into the decision making process, the component manufacturer or supplier. The component itself could constrain the design and both designer and builder will need to be aware of this. It is likely that the component manufacturer/supplier will be under contract directly to the client or builder, under the direction of the architect. In either case the manufacturer becomes an important party to the process.

There are examples of component producers determining the main contractor's requirements as regards the physical determination of the building and the letter of the contract. One such example is the structural framing system, either in steel or precast concrete. It may be that the main contractor is in overall control but the major element is the supply and erection of the frame, together with floors and roof. In essence the whole of the work is centred on the supplier and the builder and designer have little influence over detail and method. It is expected that the frame will conform to the basic design brief and that it meets structural and technical design criteria and regulations. There are companies which offer design/build systems based on components. They are then in the role of supplier and builder and enter into a prime contract with the client. This arrangement does not automatically exclude the role of architect: he may be brought in to oversee the project with regard to setting the building in its environment, selecting finishes and acting as independent adviser.

ON-SITE TECHNOLOGIES

It is on the site that component construction makes the greatest impact. Its general implications for sequencing of activities, both during design and in construction, have already been discussed. The other factors are listed below, to be discussed in greater depth in the chapter on Assembly:

range of skills for handling and fixing
plant and machinery required
site organisation function
setting out procedures
quality control procedures
ensuring good joints
site environment conditions

MAINTENANCE

A component in a motor car or washing machine can be replaced at some future time if it fails. Generally, this happens as a result of general wear and tear. A cynical view is that there is a built-in obsolescence to these types of goods so as to ensure a future market. What could be more appropriate than to make those components which are liable to wear and tear easily replaceable? This can be quite easily achieved in the case of a building's services but it is more difficult with the fabric. It has to be assumed that the structure, even if of component construction, will not require repair. Where deterioration might be expected is in the windows, doors and possibly floors. Roofs have problems of their own which may create the need for major overhaul. Doors can be easily replaced. Windows are harder to replace and are generally designed for the life of the building. The approach now taken is for the need for constant care and attention to be eliminated and the component made to last. The accent is on producing structural and fabric components which will not need replacement, rather than to produce and fix for replacement.

The position has not yet been reached, if indeed it ever will, where a section can be taken out of a building and replaced with a similar or new component.

SUMMARY

More and more items are being built into buildings in the form of components, finished and made off site. As design may be based on a selection of 'off the shelf' units the manufacturer becomes a more important element in the construction process.

QUESTIONS

1. Taking an item in a building show how it has been developed into a component.
2. Discuss the implications for the builder if he is faced with a component-based scenario for the purchase of all building items.
3. Should technical literature be improved in order to ensure its more widespread use?
4. Should the trend towards component construction be resisted?

5. Component Systems

In systems parlance there are two broad categories. An 'open' system allows for the inclusion of components and parts which are not necessarily manufactured by the initiator. A 'closed' system can only be created using purpose-made components. In between these two categories there can be a system which is reliant upon some purpose-made items but which is sufficiently flexible to take a variety of ancillary components. The high-rise structures built in the 1960s are examples of a closed system. Walls, floors, staircases and roof were totally integrated and developed and produced by the same manufacturer. They were marketed under brand names and sold as complete units, albeit with some scope for variance in finish, layout of internal rooms and height. It was not possible, without major alterations, to incorporate components from another manufacturer.

CLASP (Consortium of Local Authorities Special Project) is an example of the meeting between a closed system and a variety of additions. Initially, it was developed to cope with building on ground liable to mining subsidence. A steel frame was designed which, by using sprung lateral bracing struts, could accommodate large vertical movements in the ground below the building. This was manufactured to a basic module and section sizes and layout. External cladding was not strictly limited and internal layouts were flexible. As this was a consortium funded by a number of local authorities a relatively large number of buildings, mainly schools in the first place, constituted the market. Suppliers of other components, such as windows, were selected on performance and cost criteria. Alternatives are available and the whole system is described in a manual. This gives the common working details etc. specification and shows the range of alternate components. The system successfully survived the fluctuations in the market and the debasement of industrialised building systems. Some measure of its durability can be attributed to the following factors:

1. considerable time and effort went into the design stages to ensure its effectiveness;
2. prototypes were built and tested and monitored very carefully;
3. any problems identified by the monitoring process were

solved and modifications made to the next building (this has continued since the late 1950s);
4. variety of external appearances are possible. Internal layouts are flexible;
5. economies of scale were achieved owing to its adoption by a number of clients, not only those who had a vested interest in it.

The system can be built under licence; there is not a direct organisation which carries out the erection process. The client normally takes responsibility for this, acting on advice. The key to its successful erection lies in the manual. This is detailed enough to give those who are unfamiliar with CLASP the knowledge to achieve successful construction. Despite the apparent drawback of having to learn the system erection procedures it is still employed worldwide.

Examples of a hybrid systems include some of those based on timber frames. Here the base structure is provided by interlocking panels and upper floors and roof. Again there is a choice of layouts and external finishes.

HISTORICAL AND FUTURE DEVELOPMENTS

History is full of examples of prefabricated buildings. In its most primitive sense the nomad who carried his tent from place to place must constitute the first and most enduring example. In the UK there is documentary evidence of components (e.g. roof timbers) being produced off site and brought on to site, cut and fitted, then dismantled, transported to site and erected in place, dating from the fourteenth century. In the 1800s there was considerable activity in the innovation of prefabricated buildings (see Herbert G., *Pioneers of Prefabrication*, 1978). Examples of prefabricated structures during this period are the Crystal Palace (the building to house the Great Exhibition of 1851 in Hyde Park, London), the barracks and hospitals for the Crimea War, and many smaller buildings which were sent around the world to provide accommodation for colonists. More recent examples have been well documented and chronicled by Russell (in *Building Systems, Industrialisation and Architecture*, 1981).

Bender has put forward three possible scenarios for the manner in which buildings could be constructed. (Bender, R., *A Crack in the Rear View Mirror*, 1973.)

1. The Housing Factory

Examples of this type of production, based on the car production line, have been attempted, and some still exist. Now mainly confined to two-storey houses, there are timber frame systems and Volumetric houses. These are made in the factory and shipped to site. Volumetric houses are delivered complete with all internal services in room modules. All that is required on site are the foundations, underground service connections, and external cladding. On completion they look the same as traditionally built houses.

As factory production becomes more sophisticated room dimensions tend to be fixed by transportation requirements, rather than by internal requirements. New materials are becoming familiar, together with methods of joining them which can be made possible by advances in technology.

In the USA the housing factory is becoming part of commercial conglomerates. These companies take over the whole process, from buying the raw materials, such as timber and land, to building and maintaining. This trend constitutes the basis of the second scenario.

2. The development of large-scale systems-oriented life service industry.

This is described as a 'shelter industry'. It will have charge of all services, supplies and systems necessary for the support of community life: public utilities, housing, communication and even recreation and medical care. It means that all facilities can be co-ordinated and integrated into a total system. Why should not the lorry which delivers from the department store take away the household rubbish? Bender goes on to say that housing may be seen in overall terms, as the product of the numbers of people in the community and the activities in which they will engage, rather than the number of one, two or three bedroomed houses needed. In such an analysis the shelter industry will seek the best 'mix' and the most efficient combination of living, sleeping, storage, cooking, sanitary space and public facilities.

At present there is a multiplicity of specialisms which produce *ad hoc* solutions to problems. In the total systems approach material production, research, system design, product design, manufacture, marketing, distribution and even servicing of the finished product should be regarded as part of a single operation. It will have to consider the whole environment.

3. Appropriate technology.

The scenario presented here by Bender is in contrast to the one described above. Now the focus is on the individual and on the concept of the house being relatively self-contained. Energy can be produced for the direct use of the house by such as solar cells. Waste disposal can itself produce energy and its products be recycled on to the garden. Closed water systems need only be topped up. High technology is used in the running of the house.

The building industry would be responsible for the manufacture and distribution of components rather than for the finished house. The critical function would be the operation of an information and distribution system. The service, which would provide details of products, delivery times, interface configuration, skills required for installation, would be introduced through a network of 'building centres'. The client would bring sketch plans to the centre and they would be assessed against performance requirements and budget. A schedule of work could be produced which would indicate possible conflicts. The centre would deal with a single source of contact, thereby reducing the complexity of dealing with many suppliers. The system is directed towards the support of the individual and small builder rather than at replacing him.

These three scenarios contain examples of what is occurring here and now. The concept of factory production of industrialised building has been embraced and been found wanting in many respects. Perhaps it was the wrong materials and inadequate interfacing, together with tall structures, that created incompatability. With different materials, better jointing systems and a closer look at society's needs factory production may yet have a part to play.

The 'all-in' life system has also been tried out to some extent. Local authorities in the UK have taken responsibility for the provision of houses, schools, roads and some public utilities. This developed from eighteenth-century industrialists' schemes such as New Lanark, where factory, houses, schools and shops, with some welfare facilities, were provided on one site.

The last scenario is not yet in evidence to any extent. There are examples of the self-contained house and the use of direct energy sources. However, the idea that the client can initiate design is counter to conventional practice in the UK, where a builder typically produces an estate with a set variety of house choices. The customer chooses from this restricted choice and has little say either on the finishings. What we do have is 'one-off' specially designed houses, on their own plots. Compared to other private developments these tend to be more expensive than 'estate' houses.

McGhie ('The industrialisation of the production of building elements and components', in *The Production of the Built Environment*, proceedings of the Bartlett Internal Summer

School, 1982) has argued that all materials and nearly all processes currently used in building are industrialised or mechanised to a greater or lesser degree. Industrialisation is seen as meeting the following criteria:

standardisation of the product;
mechanisation or automation of production;
concentration of production, purchasing and marketing.

It has been predominantly carried out by manufacturers and heavy industry, and this 'primary industrialisation' leads to the industrialisation of building. Its evolution has been characterised by:

a quantitative increase in production of materials and components;

a widening of the range of materials available;

the invention of new materials, components, installations and processes corresponding to the development of science and technology and in response to widening and deepening social needs and uses, as reflected in the diversification of building types and of user specification;

a general shift in the division of labour away from 'traditional' building trades on site to industrialised production processes and mechanisation in factories;

a sharp division of labour between design and construction, and the increased importance of design in the overall building process;

a widening of the design element in building and a division of design labour into discrete 'design' professions;

'technication' of design and changes in design philosophy corresponding to the changes in the forces of production;

increased mechanisation of non-site production and of design operations;

decline of traditional skills and emergence of new and transformed skills;

increased use of systematic management techniques in construction operations and in design offices;

a relative de-skilling of site operations, with variations over different branches of production. Many new skills have come into being and take the place of existing or declining skills.

Also to be added to this list of McGhie's is the increase in tools, machinery and plant for on-site activities.

The industrialisation of building, therefore, is not limited to the systems concept. That can be considered as an offshoot from the general trend which has been gathering pace since the 1940s. There is a clear differentiation between systems building and industrialisation. Systems are generally based on one concept with a limited variety of alternatives. The industrialisation of building is an overall concept which has much substance. At present there appears to be an inexorable trend towards greater specialism based on manufacturing rather than building principles. The architect's role is one of combining the available products in design and the builder's that of assembling those products.

INNOVATION

There is a constant stream of new and modified products available for the construction of buildings. The great majority of these are initiated by manufacturers or organisations with interests in construction and in manufacturing products for buildings. Bowley (*Innovation in Building*, 1960) has shown that most innovation is undertaken by manufacturers, not by builders or designers. Since that study there has been no evidence to contradict him.

Many of the products cater for all types of work, from new build, through refurbishment, to repair. They are also available to a large variety of clients, from builder to large building owner to do-it-yourself householder. Indeed, some products marketed for DIY have been bought and used by the builder in preference to trade products. Examples are wall tile adhesives and tiles with spacing lugs and/or chamfered edges. Similarly with the use of power tools: the electric drill sold for home use, which is still very popular, is now common in the tool bag of the professional tradesmen on site.

In one way the industry is spoilt for choice, as there is now such a wide range of materials available; the right selection can prove difficult and in many instances there are no controlling bodies to ensure a consistent standard. Many products comply to a British Standard, but there are still many instances where none have been formulated. It can be many years after the introduction of a new product that a standard is agreed between interested parties. Normally, a panel is set up which comprises members of the British Standards Institution, representatives of manufacturers of like or similar products, scientists or researchers in a field relevant to that material and/or product, independent experts, practitioners or users. These standards may be limited in their scope, only covering the process of manufacture. Rarely are criteria developed which relate to the use of the product. Standards are, in general, readily accepted by manufacturers and, assuming that they are fair and reasonable, there is little opposition to their acceptance. If specifiers refer directly to a standard in their specifications and drawings, which form part of the building contract, it can be legally enforced. It is then the responsibility of the builder to ensure that the goods and products bought meet the terms of the contract. Failure to establish the credibility of the product can result in non-compliance with the terms of the contract. The least that would be expected by the client is substitution of the non-conforming goods with others seen as complying with the standard. At worst the contract could be declared void at the expense of the builder.

A body which has found some difficulty in obtaining general acceptance is the Agrément Board. It is intended to be financially self-supporting but up to 1985 was still being supported by government funds. The main problem is that relatively few manufacturers are willing to submit their products for assessment. This assessment is not based on the manufacturing process but on the performance in use of the product. Manufacturers' reluctance to become involved is understandable. In the first place submission is not obligatory, and specifiers can have little influence on its enforcement because:

(a) if there is no product with an Agrément Certificate available for a particular function nothing will be gained by specifying one;
(b) even if one or two products are available with the certificate these may be deemed inappropriate for the function;
(c) the choice will be very limited, to the disadvantage of client and designer.

As far as the manufacturer is concerned there is a problem regarding control over the product after it leaves the factory. Generally, it is delivered, handled, fixed and then used by others. The manufacturer can only specify and recommend how this is to be done. He has little influence over the actual process of construction or with respect to the component's interface with other items in the building. Adjacent components or the local spatial environment could have a detrimental affect on its performance. This could lead ultimately to reassessment and non-renewal of a certificate. This non-renewal could be more damaging to the manufacturer than if there was no certificate in the first place. A reputation sealed with a piece of paper can be harder to sustain than one built up gradually on use and experience. Of course it could be said that those manufacturers who do not offer their products for assessment must have doubts about their quality. But should technological advance be governed by the granting or otherwise of seals of approval? If these assessments are set to too low a standard then they will not be effective in protecting the client. If set too high they may drastically reduce the range of options.

PERFORMANCE OF COMPONENTS

Another approach to the achievement of acceptable standards for products and components is based on the performance concept. This has already been described in Part Two on Design. Here its implications for components will be considered. Assume that the designer has produced an overall design which sets the boundaries for the building but does not provide detailed recommendations for materials and components. What is given instead is a performance specification. By this means responsibility to meet those specified requirements is transferred to the manufacturer or supplier of the component. As will have been seen these requirements can be far-reaching and detailed. In meeting them the manufacturer may have to introduce further tests and assessments in order to prove compliance to relevant standards. It can be a rigorous method of securing fitness for purpose.

This approach need not be confined to the designer. There is nothing to prevent the builder from employing performance specifications in obtaining goods and services; so long as a practical set of criteria can be presented, then this could become a condition of the contract. Again, this puts a large measure of responsibility upon the manufacturer.

A further ramification of performance specification is that the designer and builder influence product development and innovation. By being definite and confident in the delineation of the actual requirements for a building and its components they can demand absolute solutions. These solutions may not be currently available. A prospective manufacturer will then have to modify or create a product in order to meet the particular requirements set by the performance criteria. An example could be the setting of noise reduction level between two spaces having restrictive design and production criteria precluding conventional solutions. A comprehensive application of this approach could lead to designers and builders initiating innovation, rather than being in the current position of having to take what is on offer, whether it is good, bad or indifferent.

TOLERANCES AND ACCURACY

A particular aspect of performance which has direct relevance to production is that of tolerances and the degree of accuracy of components. It is implicit in the shift from on-site building to off-site manufacturing that a high level of accuracy can be achieved in the size of components. Although this depends on the basic materials used and to what extent their nature allows the achievement of exact dimensions, a product constructed under controlled environmental conditions should have a greater expectation of meeting tolerances. Moulds can be carefully made and in materials which give consistency of reproduction from many uses, as for concrete products. Machining and adjustments can be better carried out on benches than in cramped situations on site.

As the nature of materials has been explored and manufacturing processes improved, there has been an increase in the ability to reach high levels of dimensional accuracy. In a building originating from a single factory it would be highly probable that all components would fit together accurately. This situation does not occur in practice, although some examples of partial prefabrication occur, such as Volumetric housing. What is common in practice is that many different components come from many unrelated manufacturers, each having their own sets of tolerances. In addition, these are combined on site with components and units built in situ. Whilst each may conform to its own particular set of tolerances, in combination problems can arise. A case already mentioned concerned in situ concrete frame and floors, precast concrete cladding panels and metal windows. Not only did the ranges of tolerances applicable to each vary widely, but when put together they produced joints which were difficult to seal, and they looked unattractive as they did not have a uniform width.

BRE has carried out much research into problems of tolerances and accuracy. Consultation of this is essential when designing joints and in order to realise their limitations for site practice. A handbook has been prepared which will help all those involved in production processes towards the successful achievement of tolerances and fit (Bonshor, Eldridge, *Graphical Aids for Tolerances and Fits. Handbook for manufacturers, designers and builders*, BRE report, 1974.

6. Assembly

As building becomes more reliant upon manufactured components and mechanised site activities it comes to resemble an assembly process. It is determined more by off-site technologies than on-site crafts. With the reduction of in situ work and a corresponding increase in the use of readymade components the concern of the builder centres on the effective combination of specialists and the physical jointing of units. Management becomes a process of co-ordination rather than the direction of labour.

It has been seen that components are manufactured primarily by those not directly involved with design and production. The designer only selects an appropriate component whilst the builder has to fix it. The on-site technology is one of fitting together the parts. This contrasts with a technology based on the mixing together of materials in place. On any one project it will be hard to distinguish clearly a technology based on assembly as opposed to one of making in place. There will be instances of the necessity to cut and fix, to mix and match or to make and mend. To give real examples of these extremes will perhaps make the differences clearer.

Consider a single-storey factory. If it is built with a steel frame, external cladding of profile steel sheeting to roof and walls (bolted together), metal windows, with internal partitions in timber stud and plasterboard, it is primarily an assembly process. The frame would be bolted together requiring no cutting or shaping; the cladding would require the minimum of cutting and shaping; and the windows could be delivered already glazed. The internal partitions would require cutting and fitting, but from sheets of plasterboard.

A brick-built structure with timber flat roof would not be considered an assembly process building. Although there will be examples of components that are put together, the basic process is that of placing, cutting and mixing to build the walls and roof. If the internal partitions were brick or block this would be a continuation of the same building process. Here we use the word building to show that the making of the structure is primarily an on-site craft activity. Assembly is off-site manufacture with on-site joining.

With component technology (see Fig. 6.3) the manufacturer is an integral part of the process and can be involved at a very early stage in the design. Whether the contractual arrangement is one of management fee or of competitive tender based on full documentation, has little bearing on the degree of involvement of the manufacturer. In either case the product may be chosen at an early stage because its identity must be known in order that the technical details can be shown on the drawings. These details need to be considered in relation to the product's enclosing/adjacent materials and components, so that they can be integrated into the design. An analogy is that of the jigsaw, but one where the pieces have been designed and produced by different people. It can be seen that in that case a problem will arise in the fitting together of the pieces. The overall jigsaw must be devised; its boundaries; its picture; its dimensions; its basic materials; the shape and size of its pieces. The rules and common means of control need to be set so that production of the jigsaw can be achieved with the minimum of trouble.

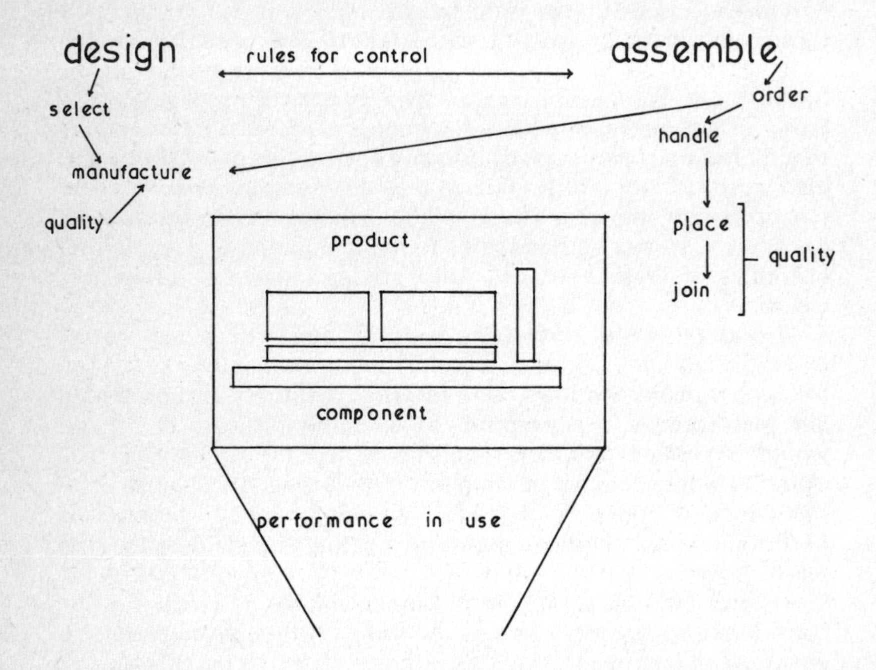

6.3 Technology of component construction

ASSEMBLY CONTROL PROCEDURES

The control procedure must commence at the design stage, whichever method is used. The procedure can be centred on a number of modes:

1. Structured sets of drawings
2. Performance specifications
3. British Standards and Codes of Practice
4. Dimensional co-ordination
5. Modular co-ordination

Each of these will, in its own way, have an effect on the process of production.

1. Structured sets of drawings

These are based on separate drawings giving the location, the component details and instructions for their assembly, plus schedules if necessary. (See BRE, 'A critical look at working drawings'). When used they can show from three perspectives the construction detail. For this detail to be effectively produced all the information needs to be at hand. This could be cumbersome as it may require up to five drawings to show the one detail in its entirety. The assembly strategy must be concerned with the integration of information in the first instance, followed by the integration of the components.

2. Performance specifications

These give a large degree of flexibility to the producer. As long as the performance criteria are met in the selection of the components and they are fitted together accurately, it is for the builder to determine the most appropriate methods of construction. It is possible that where no suitable components are available then built-in-place methods will need to be applied. The onus is on the builder, together with suppliers, to ensure that the components meet requirements. The duty of care is explicit.

3. British Standards and Codes of Practice

These are commonly used on all building work of any worth. They lay down minimum standards for materials and components and recommend methods of construction. The codes of practice are applicable to the builder for the work on site. They are based on experience and practice and take into consideration such factors as safe working procedures. They can become dated as new materials and procedures are introduced, with the consequence that they may then be inappropriate to the task. The Building Regulations, Approved Documents use British Standards and Codes of Practice extensively as the means to meet requirements.

4. Dimensional co-ordination

By superimposing grids, zones and reference dimensions the designer can indicate a framework for the integration of the components. Firm control lines can be installed to give a real point of reference. Into these reference points the components can be fitted. It means that the correct placement and integration relies on a system which has to be understood in itself before application. When learnt and operated satisfactorily it does provide a good basis for the achievement of sound assembly.

5. Modular co-ordination

This is dimensional co-ordination structured around a set of dimensions, each a multiple of another. It presupposes that components and elements used will also be based on the same set of dimensions. Once understood, the system need be no more complicated than ordinary dimensional co-ordination. Its main implications are that components have to be manufactured according to the module. This may limit choice of component and restrict methods of joining. In theory however it could lead to assembly processes which, after time for learning, are simple, quick and economic, with the added advantage of constant quality levels.

In essence, any form of assembly control has to be centred not on components but on the gaps between them, as shown in Fig. 6.4. The major problem for the site assembly process is the accomplishment of joints with integrity.

It seems that many problems of human organisation and its products revolve around aspects of integration. In the physical sense this is effected by means of joints between components; in the managerial sense through continuity and communication amongst specialists. This can be illustrated by the analysis 'Faster building for industry' (NEDO, 1983) from which an analogy may be drawn with the process of jointing between components. The survey compared time, cost and quality with respect to four principal methods of contractual arrangements for construction, namely: traditional (main contractor with sub-contractor); own management (client organising process); design and build; separate management function (project management). The general conclusions were that contracts organised under the separate management function showed a significantly higher proportion, 50%, being completed in faster than average times, with no increase in cost and no deterioration in quality. This compared with 17% for the traditional mode. 12% of separate management function projects were slower than average, with 46% of traditional being slower. The most common cause of hold-up or delay lay in sub-contracting, affecting 49% of the projects studied. This was distributed equally between nominated and domestic sub-contractors, but was often associated with other factors. Delays earlier in the programme would necessitate a change in programme which sub-contractors found themselves unable to accommodate in revised programme dates. Subletting by contractors also led to similar delays, amounting to 6%. It appears that the problem of delay was not caused by dilatory performance once on

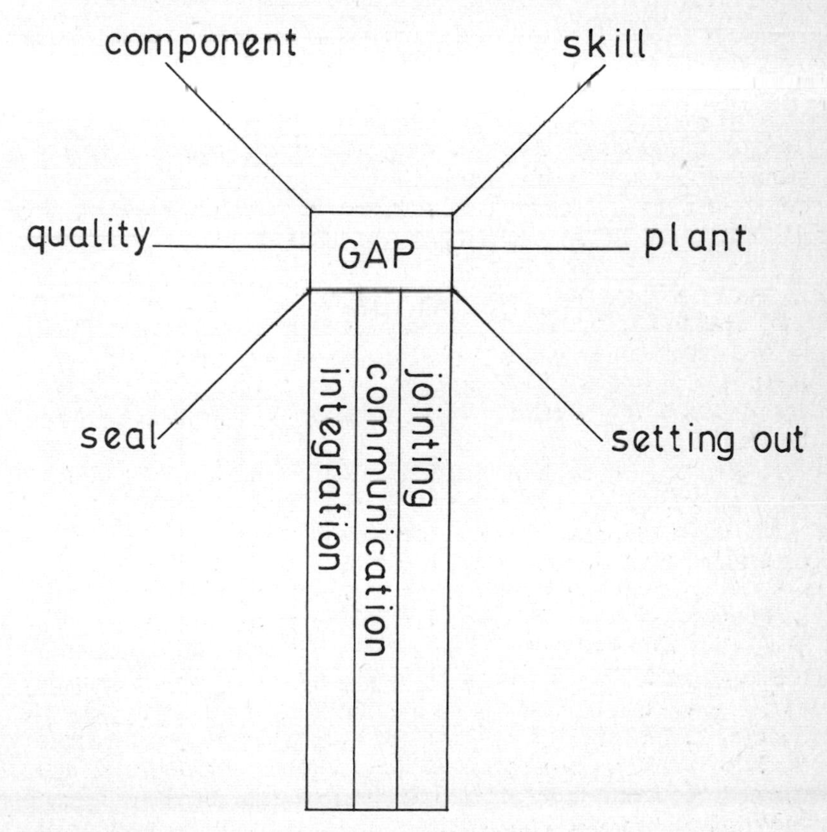

6.4 Assembly

site, but that it was difficult to achieve a continuous flow of work. This can be seen from Fig. 6.5. A delay in activity 1 has caused a gap to develop before the start of activity 2. As a consequence the overlapping relationships of activities 3 and 4 are now shortened, causing further overall delay. In other words it is not the activity of the specialist which causes the problem – they achieve their performance requirements – it is the process of closing the gaps.

Similarly it can be seen that the component figure in the same diagram has gaps which need to be closed. The key to successful component assembly lies in the ability to obtain joints which are easy to complete. This ease of completion take into consideration the following factors:

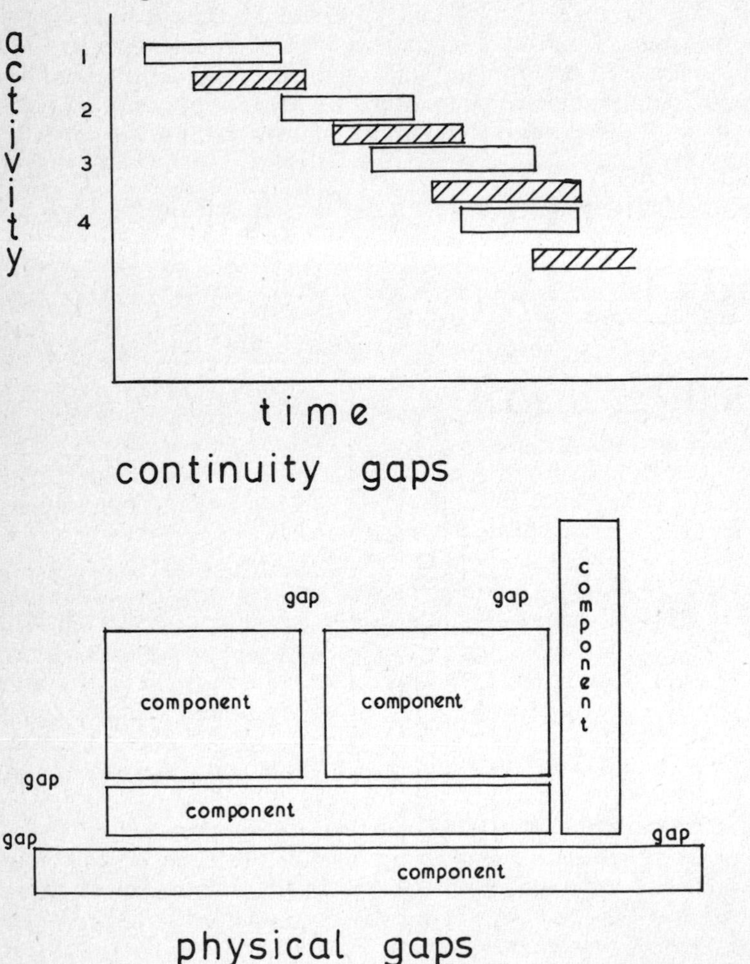

6.5 The gap

(a) Relative material/component characteristics
Here the concern is with compatibility between the adjoining component/materials. Will any problems arise due to induced corrosion? Will differences in relative movements disturb the joint's equilibrium? Can the joint filler successfully cope with the different qualities of the materials? These are termed inherent deviations.

(b) Size of joint
The optimum size of joint must be determined, using the following criteria:

tolerances – component size
– movement allowance
– fixing procedures
type of joint filler or cover.

These are termed 'induced variations' and are the result of human inability to obtain absolute precision.

(c) Economics
Allied to the above two factors is the cost of achieving a successful joint. If there is a human inability to obtain absolute precision, how far can one go in this quest? The limiting factors will be time and money. Dalton and Renward have argued that it is necessary to consider the costs of components as related to manufacture, but to bring into the reckoning the benefits and penalties which their use imposes on the other stages of the construction process (J. Dalton, K. Kenward, 'Economics of dimensional accuracy' in D.A. Turin, ed., *Aspects of the Economics of Construction*). If a manufacturing quality control procedure is to be used, each item would need to be inspected, i.e. a 100% sample. This cost would normally be prohibitive. If the reverse action is taken, i.e. no checking at all, it is likely that the cost of on-site rejections would become prohibitive. A balance must be struck in reaching tolerances which are acceptable to both manufacturer and builder. As on-site fitting and adapting can have major implications for the sequencing, timing and cost of the construction process, the bias must be towards getting the right tolerances in the manufacturing process. If it is uneconomic to carry out 100% checking in the factory, this must also be true for the builder checking all items on arrival on site.

Unfortunately, little is known about the cost of achieving accuracy in setting out and fixing on site. Problems arising from inaccuracies are hidden by the confusion of interests of those parties involved, and by the need for completion on time. Also, what might be acceptable as a level of accuracy on one job may not be acceptable on another. Owing to the great variety of permutations for site-fixing details and jointing arrangements it is impossible to arrive at a once-for-all optimum level of accuracy, applied either on or off site.

It must be recognised that inaccuracies will occur and that the endeavour should be directed towards the economic absorption of these, whilst still meeting the performance requirements. As Bonshor and Eldridge have said, 'joints are interruptions in the physical continuity of an assembly's functions' (R.B. Bonshor, L. Eldridge, *Tolerances and Fits for Standard Building Components*, BRE CP 65/74).

In the conclusion to their paper Dalton and Kenward state:

> What is good for the manufacturer may not be good for the sub-contractor who fixes; low costs in manufacture and erection of a particular component may result in extra costs in jointing or other components; rejection of components on site will lead to delays in construction and extra costs to the contractor, and completion with a cost penalty to the client. In developing, manufacturing and using components, particularly for large building programmes, it is essential that the interests of all parties are considered and that the sums be done at the right time.

The techniques of jointing must be fully considered in the construction process as each will influence the process. Joints and fixings can be classified under the following heads:

Flexible
Where a jointing material is used which can accommodate a relatively wide range of movement and tolerances. An example would be a mastic type of filler. A fixing can also be flexible in that it allows a tolerance for fixing or subsequent movement; here an example would be an oval slot for a holding bolt.

Solid
Where a rigid joint is required between components to give structural integrity, for example plates and bolts joining steel beam sections.

Cover
Where it is impracticable to achieve a joint or fixing which can accommodate possible movement then a cover or third element needs to be introduced. It is often used in the jointing of metal window frames where a clip-on moulding spans the connection between the two units.

A flexible joint can be applied during or after the fixing operation, whereas a solid joint has to be done at the time of erection. The third element joint is generally carried out simultaneously with the fixing, after checking for alignment, security, etc. Some flexible jointing techniques are considered as a separate activity and, consequently, are carried out subsequent to the fixing of components, by another trade or by a return site visit.

CONTRACTUAL RELATIONSHIPS – EFFECT ON ASSEMBLY

It has been stated that, irrespective of the contractual relationships, there is a need to consider the technical details of components at an earlier stage in the design and construction process. This still holds true, but with the increase in the use of management fee type of arrangements there are some further ramifications for assembly technologies.

The basis of management fee is that the builder organises, procures and controls the work of specialist contractors on behalf of the client, for which a sum (fee) of money is paid. All the major activities of the contract are parcelled up as packages, for example: the structure and floors; the external cladding; the internal partitioning. These are let as separate contracts, usually on a normal competitive basis. The contract is between client and specialist. The designer/architect may be under the full control of the management contractor or independently employed by the client. In either case the designer is not responsible for many of the details of construction solutions and processes: these are the responsibility of the specialist contractors. They are required to show how their component, material or item of work meets the overall design criteria to provide the contract information for the technology employed. This is integrated into the building's design by the architect and the management contractor. No prime bills of quantity are produced; it is the responsibility of each specialist to provide his own quantities. The extent they go into the preparation of quantities is their own concern, so long as they are sure that the work package can be completed to the given estimate, to the overall design and to time.

Instead of the designer dictating the full detail of construction to the specialist, the specialist now has the opportunity of influencing the technology. Where the specialist is experienced and confident there will be a willingness to participate in the design and build planning stages. Many benefits could accrue to the specialist, such as ensuring that his particular requirements and constraints are met; he will have foreknowledge of the possible problems at an early stage, he can integrate fully with the other specialists, and all these factors will lead to greater efficiency and productivity with their subsequent monetary rewards. On the other hand those specialists who are used to being 'spoon fed' by the architect and builder via detailed drawings and bills of quantities and plans, and who rely upon these to estimate their work, will find the new expectations disturbing. As many management contractors have said, in their initial dealings with prospective specialists they have had to educate them to the need to provide much more detailed information. In some instances this has been a long and painful process.

As the role of specialist contractors is expanded they should exert a major influence over design and build decisions. They will be looked upon as experts and as such will be able to provide the relevant information and work performance. They will have the opportunity to suggest alternatives and introduce innovations. Under some management fee contracts the client can reward the achievement of savings. Where a builder as manager, or a specialist, produces a saving on the initial cost estimates by modifying the design, then the monies saved can be shared by client and contractor. With a traditional contract and sub-contract it is not in the financial interest of all parties to produce savings, as these will only benefit the client. Additionally, the overall cost will fall and therefore the total profit will fall, if rated as a percentage of the contract sum. It could be said that traditional contract arrangements, with a detailed design and bill of quantities, lend themself to emphasis on money at the expense of technology. Where competitive prices are tight contractors will seek to find omissions in the bills and extra works in order to increase the contract sum and thereby swell profits. A close look at the technology to seek savings or increase profitability is rarely undertaken, as most builders and specialists consider they cannot influence the designer, especially after the preparation of the drawings and the bill of quantities.

It will be interesting to see if there are any measurable improvements in technology as a result of the management fee type of contract.

HUMAN FACTORS IN THE ASSEMBLY PROCESS

As building becomes more industrialised, i.e. as components reach site in a finished state and are assembled with the help of plant and machinery, there needs to be a corresponding advance in knowledge of the processes of technology. Such knowledge will show where improvements can be made and efficiency increased. It has already been seen that the science of human factors can aid the design of buildings: the same principle can be applied to the construction process. Remember, one is looking at the widest aspects of ergonomics, as they are integrated into a systems approach to analysing and predicting behaviour.

Even a relatively simple operation requires a number of prior steps to be completed before it can itself be executed. Therefore each operation cannot be considered in isolation. From first to last the process should be regarded as an interrelationship of human behaviour within the structure of the technology, affected by the psychological and physiological characteristics of the tasks. The fixing of a metal cladding panel to a steel frame offers an illustration. In order to do this the worker is reliant upon a number of ergonomically determined factors:

(a) *access* – the position of the panel and its fixing nuts and bolts have to be within the reach of the operative.

(b) *handling* – the panel itself must be within the capability of the operative. It may require a second person for lifting or positioning, or some lifting equipment. The equipment must be capable of holding the panel securely and allowing manoeuvrability into position. Will it need to be moved for subsequent panels?

(c) *fixing* – will holes have to be drilled in the panel? If so what sort of drill will be necessary? Can it be operated and moved by one person? Can the holes be drilled when the panel is in position? Will temporary fixings or supports be required? How will efficiency be affected by drilling in place, or the fixing stage? How are the fixings made? Is it necessary to produce a set torque? Can the fixings be completed from one side only?

(d) *jointing* – here the ease, or otherwise, of the operation can be critical. What degree of accuracy is demanded? Will it be necessary to cut and fit? If so, what tools can be used and in what position is it best carried out?

(e) *assembly* – this is probably dependent on tools powered by an external energy source, as opposed to hand power. The energy source has to be located in such a way that ease of movement is not restricted in reaching all the panels. Will sub-distribution points be required?

(f) *support* – prior to the fixing of the panels is the activity of securing their supports. What level of accuracy is required here? What are the repercussions of poor quality on the subsequent panel fixings? How difficult or arduous is it to ensure that the panel bearers are successfully positioned?

(g) *maintenance* – when the panels are fixed and in use can they be maintained effectively? Will subsequent repair and maintenance depend primarily on the success of the on-site operations? Does the joining integrity depend solely on the level of care and attention applied during the assembly process?

Although the wealth of experience and practice in carrying out the majority of site assembly operations is not in question, there is little published analysis of these in terms of human factors. Whilst manufacturers of power tools ensure that their tools are designed ergonomically, this concern is rarely translated into the assembly process on site. Work study methods are one way of looking at particular problems, but even when applied they fall short of relating an activity to others preceding and following.

There is considerable scope for the builder to apply some of the principles and techniques associated with the science of human factors to construction activities. In so doing, he will at least gain a greater understanding of what is actually going on, and at best will highlight problems which need attention. The art of predicting human behaviour in the context of technologies in the construction environment can, then, be used to benefit operative, builder, designer and client.

Much more can also be done by way of analysing the relationships between men and machines, where machines are at the central point of an activity. In the case of the tower crane, for example, some studies have been carried out which throw light on its capabilities, efficiency and potential integration into the overall site assembly system. One can also find examples of plant manufacturers perceiving a need and then producing a piece of equipment to satisfy that need. One such is the introduction of small lightweight wheeled tractors with front bucket. These are small enough to travel into buildings with average floor to ceiling heights, to be lifted up to higher floors, and to travel upon these. They can be used to clear rubble and waste, carry materials and components. They are ideal for refurbishment jobs where internal demolition is necessary. When these machines are used the sequencing of the activities and the employment of appropriate technologies can vary from the traditional practice of using human labour to move waste and materials. Larger components can be carried out quickly, giving complete access for following trades. Now that a machine is used a different skill is required, that of driving and operating. What training is required? Is there a need for a banksman? Can the waste material be discharged and cleared from the building at a rate fast enough to meet the output of the machine? What temporary works will be required to cater for the machines?

The machine becomes an integral part of the system and its advantages and limitations an extension of the operative's capabilities. It must serve the operative as well as the task. If it is uncomfortable or awkward to use its effectiveness is lessened, together with its practical and economic utility.

The introduction or application of new processes, skills or machines should be analysed with respect to human factors criteria to ensure their effective and safe assimilation into the assembly process.

AIDS TO CONTROL

We can use physical aids to help in the correct setting out of the components, but this only provides a framework for assembly. The use of such techniques as dimensional or modular co-ordination, grids, centre lines or zones can indicate, from the drawings, positioning and tolerances. What is further required is an attitude that accepts the need for high levels of accuracy, and care and attention to details of construction. Although the headquarters of the Hong Kong and Shanghai Banking Corporation building in Hong Kong can be seen as a unique design, it points the way to future building processes. It is a building that is designed and constructed to precision engineering principles. For example, the suspension trusses off the main support towers have been linked with giant cotter pins. To achieve this level of accuracy the engineers had to use sophisticated measuring and positioning equipment. This, in turn, ensured that the subsequent cladding and components could be accurately manufactured, in the certainty that as construction progressed each component would fit exactly. Some of these components were room-sized fully fitted service modules which were lifted into position. The acceptance of precision engineering standards is the key to effective control on site.

In adopting any means of control it is essential that all those using the system are familiar with it. The system must not be so complicated as to need a major training programme in order to learn it. Nor should the physical site lines and datums result in a spider's web which hinders production. From the designer, to the site manager, to the subcontractor, to the operative, all should be aware of the system of control. The effect that attitudes have on the technology of building is explored in the BRE study, 'Quality Control on Building Sites' (M.J.C. Bentley CP 7/81). This shows that achievement of the required quality standards was dependent on the level of interaction on site between architects, clerks of work and contractors. Sites which showed good quality levels were generally marked by a commitment to the job on the part of everyone concerned: 'Everyone becomes involved to some degree in dealing with and helping to sort out problems, not necessarily to do with their own work, but involving themselves with the job as a whole.' The study highlights one site as exemplifying poor quality standards:

> On this site there was a lack of effective management coupled with extremely poor project information (often incomplete and sometimes incomprehensible). Everyone was performing strictly within the limitations of their formal role. The clerk of works was dealing with problems alone, with little consultation with trades foreman or the agent. The architect was reluctant to visit site. When he did he was inundated with queries arising from the drawings. The clerk of works was unhappy with the level of quality on the job but was unable to do anything about it. Lack of care by both contractor and architect was proving too much for him to handle.

The author's main conclusion was that standards of quality do not rely on formal checking and acceptance or rejection of completed work, on the site agent and clerk of works putting great effort into creating an environment where good work could and was likely to take place. In recapping, it is not the stated control procedures which determine the levels of quality or site technologies but the attitude of those engaged in the assembly process.

ENERGY REQUIREMENTS FOR ASSEMBLY OPERATIONS

We have seen how the building industry is becoming a highly industrialised activity. More and more units are made off site in factories, which in themselves are geared to sophisticated production techniques. Similarly on the site the use of plant and machinery has developed significantly. However, the human being's energy derived from food is now replaced by power which has to be generated. Where once we had a dozen men digging foundations we now have one backacter which can do the work in half the time; but we have also to provide fuel for this machine.

There are two main sources of power for the operation of site machinery: electricity and diesel. Gas is used on site but mainly for space heating or direct heating in the form of blow torches. This is supplied in the form of cyclinders which require safe handling and storage procedures.

Electricity on site can be produced from the mains or generated from independent units. In the former case it is advisable to reduce to 110 volts for safety, but for some heavy capacity plant a three-phase supply might be required. We have to be careful about the means of distributing this around the site. Decisions need to be made, which will affect the building technology, on what type of electricity to use. Can we get cables to the place of work? Can a generator be used which will not foul the atmosphere? We will need another supply of fuel to operate a generator: how do we ensure its safe handling, storage and distribution? What does this power cost? Will it increase productivity?

The days of diesel as an energy source are likely to be numbered. The oil fields are finite and we can be sure that sooner or later their reserves will run out. The construction industry, amongst others, will soon have to consider alternatives to the use of fossil fuels to power digging and moving plant. There are two possible developments: (a) that we go back to using many people to dig and move materials (this would be cheap in energy but might be expensive in monetary terms); or (b) that another power source could be developed for the plant, such as electric motors from a battery supply. At present there are two problems in the way of the latter, the first being the efficient storage of electricity in the batteries, and the second the achievement of the necessary power output required to move heavy materials. Although these may not be seen as the builders' problem, since they do not make the machinery, as users they will be most concerned if an alternative energy supply is not forthcoming. Building technology will be greatly affected. Therefore the builder must monitor progress in these areas in order to be able to predict the technologies required for the future.

SUMMARY

In this chapter we have looked at the changing nature of the building process. It is becoming more of an assembly operation, using prefabricated materials and units, than an in situ building process. In order to carry out this type of process successfully we need to have adequate control procedures. The biggest problem in any assembly operation is ensuring that the gaps between the components are bridged adequately. This must be done to an economic cost and we raised some of the issues regarding it. It may be that some contractual relationships will have an influence over the technologies and we saw how one, management fee, might bring about a closer investigation of the building technologies instead of looking at alternatives purely in monetary terms. As human factors thinking can set out a system for the design process so can it also be used to improve site assembly practices. The aids to control, and control procedures in themselves, cannot ensure good assembly practice: this depends very much on the attitudes of the people engaged in the process. In many instances an informal system reaps the better rewards. Finally we had a brief look at the energy requirements for building work and saw that they have increased but that in the future some changes are likely.

QUESTIONS

1. Describe a building which has been assembled rather than built.
2. Discuss how the use of one of the assembly control procedures above would directly influence a technological detail.
3. Discuss how the quality of joints needed between components can be ascertained and achieved, commencing from the design stage.
4. Should we strive for the absolute in the quest for accuracy, whatever the cost?

5. Describe how the design/build type of contract can affect the assembly process.
6. Take a particular site technology and show how the application of human factors analysis could lead to improvements.
7. Should technology be concerned with the attitudes of people towards it or is this the province of management studies?
8. To what extent should the builder be concerned with energy supplies and consumption for site activities.

Part Four MAINTENANCE

1. Introduction

What is maintenance? If it is defined as the preservation of the existing building stock, should this be maintained to its original state, or to current standards? Why is maintenance necessary? Is it inevitable or can it be reduced? Are there preferred ways in which maintenance tasks can be carried out? If so, which give the best results to the client?

In this section these questions will be considered and some answers provided. Attention will be given to the value of maintenance, to the effect of initial design decisions, the buildings as consumers of energy, and to approaches and strategies which will aid optimum maintenance solutions.

If buildings are regarded as an assembly of components and parts, then it follows that these may be replaced; what are the implications for maintenance? There must come a time when, for most buildings, it is either impractical, disfunctional, or too costly to continue to maintain. What are the alternatives? The cyclical notion of the use of land and buildings will be discussed and the influence of the computer in forecasting and controlling maintenance works will be assessed.

BS 3811: 1964 defines maintenance as 'work undertaken in order to keep or restore every facility, i.e. every part of a site, building and contents to an acceptable standard'. The question of acceptable standards rests, in most cases, with the owners/users of the building. It is possible to maintain some items to a level above that originally achieved. This may be due to present standards and qualities being higher, or different from those at the time of the original construction. So whilst the item may have been maintained it might also have been improved. The definition indicates two broad strategies for maintenance. In 'keeping', work is carried out in anticipation of failure (preventive maintenance), whilst in 'restoring' a failed item is replaced (corrective maintenance).

Coupled with the notion of maintenance are the processes associated with conservation, rehabilitation, preservation and adaptation. Conservation is generally applied to the maintenance of older, perhaps unique buildings, which have some architectural or social significance. They may be listed buildings and buildings with preservation orders. The idea is that they should be kept in their original state. Any maintenance work must be carried out using like materials where possible and in the original style.

Rehabilitation can be very wide ranging. Some buildings are completely gutted, leaving just the facade (or some parts of it) and a new structure built behind. This is one form of rehabilitation. Another interpretation of rehabilitation is where internal services and accommodation not meeting present-day standards are stripped out and replaced by new, accommodation altered and the fabric treated to give a further substantial lifespan to the building. This approach is commonplace for housing.

Many building clients establish the level of maintenance provision on economic criteria. A certain budget is set aside annually within which all tasks have to be put in order of priority, in the case of conflict or where the budget is inadequate. Rarely does a building owner set out standards of performance, function and appearance upon which a maintenance budget can be based. Maintenance is determined by economic criteria, not by building/user/performance criteria. It is idealistic to expect monetary considerations to be subservient to functional and performance requirements and it is unknown for maintenance programmes to be drawn up without regard to the cost implications. Many building owners, or those managing maintenance, base their requirements on last year's budget, with an allowance for inflation and taking into consideration any further properties acquired. Little thought is given to the fact that as the building gets older the likelihood of additional maintenance is high, over and above that already carried out.

Economics in the widest sense also affects the maintenance of buildings. If the owners/users are experiencing financial problems, maintenance is one of the first costs to be cut. The UK government's policy of cutting public spending in the 1980s has affected local authority maintenance budgets. In spring 1985 it was estimated that £19 billion was required to bring existing local authority dwellings up to present-day standards (Building Employers Confederation report 'Spotlight on Housing Maintenance and Improvement', BEC 1985.) Other government statistics have shown that there has also been a marked increase in the levels of maintenance required on dwellings in the private sector, £2,500 (1984 prices) needing to be spent on 4 million properties to bring them back to acceptable standards.

In conjunction with economic factors, which are liable to change, is the legislation which lays down minimum standards for buildings and their maintenance. Such legislation includes the Housing Acts 1957, 1969; Public Health Acts 1936, 1961; Factories Act 1961; Defective Premises Act 1972; Health and Welfare at Work Act 1974; Offices, Shops and Railway Premises Act 1963; Fire Precautions Act 1971. In a number of these are specific references to the building being 'properly maintained'. In one act this phrase is interpreted as 'maintained in an efficient state, in efficient working order, and in good repair' (Factories Act 1961).

To give a particular example of legislation affecting standards and levels of upkeep one can refer to the 1969 Housing Act, which states that where an application for an improvement grant is made the following requirements must be met:

(a) good state of repair and substantially free from damp. This includes stability and applies to all parts of the building. It goes beyond the public health concept underlying the fitness standard, and implies that the dwelling is up to present day standards;

(b) each room properly lighted and ventilated;

(c) adequate supply of wholesome water laid on within the building;

(d) efficient and adequate means of supplying hot water for domestic purposes;

(e) inernal wc, if practical, otherwise readily accessible outside wc;

(f) a fixed bath or shower in a bathroom;

(g) a sink with suitable arrangements for disposal of waste;

(h) a proper drainage system;

(i) each room with adequate points for gas or electricity;

(j) adequate facilities for heating;

(k) satisfactory facilities for storing, preparing and cooking food;

(l) proper provision for the storage of fuel (where required).

There is no recommendation for the general state of the decorations but the above list is comprehensive and is related to present-day expectations.

SUMMARY

In this introduction the value of maintenance work that needs to be done on buildings to keep them in an acceptable state of repair has been highlighted. A reminder has been given of the influence of conceptual framework factors on maintenance.

QUESTIONS

1. How would you define maintenance? Is it different when seen from the point of view of (a) the building owner/user (b) the building maintenance organisation?
2. Is it possible to differentiate between maintenance for public and for private buildings? Is one more important than another?
3. Are there any national policies regarding the need for maintenance?

2. Value of Building Maintenance

The value of building maintenance is not easily recognised by those who see it as a drain on resources. It is seen as something which in most cases can be put off to another time. It need not be carried out until a failure occurs. In buildings, defects are tolerated for much longer than they are in motor cars, safety aspects notwithstanding. A telling comparison is that of the amount of time spent by the average motorist on cleaning his car with the average time spent on cleaning the external paintwork of his house.

Noble ('The value of building maintenance', V. Noble, CIOB Maintenance Information Service, paper no. 13, 1980) has identified value under the following headings

(a) volume and quality of work carried out;
(b) value of the things and processes accommodated in the buildings;
(c) resale or rental value of the property.

Volume and quality of work carried out

Maintenance can be regarded as a percentage of the total owners/users income and/or expenditure, on a yearly basis. As the years go by the amount spent accumulates and can finally reach 50% to 60% of the total monies attributable to the building, including capital costs. In other words the overall costs can equal the initial cost. Whether the building is under the care of a large commercial organisation or a private individual, the amount of money spent on its maintenance, as compared to annual turnover, can be as low as one per cent. When running costs, rates etc. are included the amount can increase dramatically. Those readers who maintain their own property (whether rented or not) should try carrying out a simple sum: express the expenditure on maintenance as a percentage of yearly income – it is likely to be quite low.

Those organisations responsible for a large estate, e.g. hospital boards, local government, large retailing companies, can have maintenance budgets of millions of pounds per annum. The control and organisation of their activities becomes a business in its own right, albeit one that spends money rather than earning it. By its very size it can have a value of significance in the overall achievement of the parent organisation's goals. For example, hospital maintenance is imperative to the efficient functioning of the fabric and of the facilities for care of the patients, for twenty-four hours of the day, every day of the year.

Alongside volume the quality of maintenance must be considered. Quality may be analysed within other criteria:

(i) standards of work carried out, skills necessary;
(ii) levels of efficiency/speed in meeting the maintenance requirement;
(iii) technical competence;
(iv) whether it just replaces to the original standard or is an improvement.

In (i) the work can be carried out to the minimum levels commensurate with the client's requirements. It may be that a 'patch' is deemed to be all that is necessary, rather than rectification of the whole area. In this case the job could be done by a general maintenance operative, whereas the rectification of a large area may need to be carried out by specialists. This does not imply that a generalist produces a job of inferior quality to that of a specialist, but that in the nature of the activity a different outcome is expected. Within the task specification a 'patch' may be done to a very high level of quality. A make-do-and-mend approach requires a level of quality which simply ensures the continuation of the function. A comprehensive approach would generally seek to achieve an overall improvement, with a longer life span. Unfortunately, it may be that the skills required to produce satisfactory maintenance are not available at that time and place, for a variety of reasons. The reasons would include:

– general national/local scarcity of those skills;
– required skills deemed too costly;
– emergency nature of maintenance activity necessitates prompt use of available skills.

The quality of maintenance can also be seen in the level of services offered in providing the remedial work. Is there a prompt response to the maintenance requirement? Is the work carried out effectively and efficiently within the budgetary constraints? Are the managers of the organisation competent in their diagnosis and solution of the defects? A large measure of value may be placed on the ability to respond, rather than on the ability to produce a lasting remedy. (Some commercial maintenance organisations advertise a twenty-four hour service and prompt attention to calls.) In this case operatives could be on constant standby ready for any task. The cost implications may be high, but if the losses incurred due to non-performance of the building or its services are much higher, the standby costs can be regarded as the equivalent to insurance premiums – something to be paid for, but hopefully not claimed against.

A value ought to be placed on the quality of the maintenance organisation's competence in diagnosing and subsequently producing acceptable solutions to problems. The skills and experience of the managers are, perhaps, more important than those used in the actual execution of the work. Whatever work is carried out it must be correct for the problem. The technology must be appropriate.

Finally, the value of the maintained item could, in fact, be increased. It is often the case that a remedy may be such that the original is enhanced to present-day standards. A defective item, or one that has constantly performed poorly, may on being rectified allow an enhanced use to be made of that function. This could lead to a greater value being given. Even in the replacement of the original function a certain value will have been maintained. To ascertain this the remedial work must be measured against the failure of that item. In other words a value must be set for the continuance of the performance function of the item. A replacement at a higher quality may have little value benefits, other than, say, aesthetic. Here it is important to measure value subjectively and in relation to the expectations of the client. For example, the quality of wall decoration can be much higher than the original; it may not extend the periods between redecoration, but for the users and/or clients the higher quality can only be appreciated.

(b) Value of the things and processes accommodated in the buildings

The maintenance of a building can be seen in direct relationship to the things or processes accommodated within it. To take a simple example, an art gallery or museum will accommodate priceless objects. If part of the building fails, say, by letting in water which could cause damage, then the value in getting a repair carried out is extremely high.

A building accommodating a production process can have a value only perhaps, in relation to that process. The building is of secondary importance, its reason for existence being to house that production process. In the event of a building failure causing a disruption in the production process, then a high value must be placed on the remedial action. In the case of a hospital a high value is given to maintenance because of the possible consequences for patients' health in the event of failures in the services.

If a building experiences a change of use, the levels and quality required of subsequent maintenance can also change. Take the case of London's Covent Garden Market buildings. Their use as a fruit and vegetable market demanded a relatively low level of maintenance. The traders themselves carried out some maintenance to their stalls. Now that the market halls have been converted into shops, cafés and public areas the level of maintenance has changed dramatically. The external image expressed in decoration and appearance, the maintenance of safe public passage and the need to attract the public has demanded a much higher level of maintenance. The value of maintenance is not only directly related to the provision of existing activities but also to their continuing attractiveness to customers. Conversely, change of use of a building from a high level of activities to a low one may carry with it a decreasing value for its maintenance.

(c) The resale or rental value of the property

In the early and middle years of a building's life its market value may depend to a great extent upon the level of maintenance it has enjoyed, discounting the intrinsic value of the land and its potential, and assuming that the building is viable as it stands. A building kept in good repair is bound to be more attractive to a prospective user than one that is obviously neglected. In the former case the highest market values are likely to be obtained, while with a neglected building a prospective client will assume that work needs to be done, costing extra time and money. Also a general air of neglect may be the outward sign of major problems caused by insufficient maintenance.

A balance must be struck between the levels of maintenance necessary to ensure efficient functioning and those related to keeping up the market value of the building. In reality they ought to be two sides of the same coin. It could prove unnecessarily expensive to introduce high levels of maintenance with enhancing the market value as the main justification. The market value is based largely on other factors, relating to supply and demand, position, use etc. of the building. Maintenance, by itself, cannot increase its value over and above that dictated by the market factors. What it can do is sustain the whole over a longer period of time. In the case of a building with a relatively high level of occupancy turnover this extension of time can be of benefit to the client, always assuming that the costs of maintenance do not exceed any increase in revenue accruing from it.

The prime objective of maintenance must be to ensure that the building functions, but in so doing it is possible to increase its value as well. Some aspects of maintenance are able to give an added value to the building, by way of extending its life beyond that normally expected.

SUMMARY

It has been shown that value in maintenance can be expressed in terms of the amount of work carried out, whether as national or commercial or individual worth and expenditure. Maintenance may be done solely to protect the functional use of a building, that is for what it contains; or 'in the extension of a building's life by effective maintenance, the market value may be enhanced. Value, as already discussed in Part Two in the case of design, must be seen from the perspective of the participants in maintenance activities.

QUESTIONS

1. Consider where, in a building with which you are familiar, some further maintenance could give added value.
2. Discuss the merits, for an organisation maintaining a large housing estate, of its having its own multi-skilled maintenance workforce.
3. Maintenance is largely a labour intensive activity. Discuss.
4. Approximately 50% of the total monies spent on building construction in the UK can be attributed to the maintenance of buildings. Is this of benefit to the industry as a whole, or could some of it be better used in new-build work?

3. Designing for Maintenance

The key to the levels and range of maintenance required in a building lies in the original design stage. Good, bad or indifferent solutions will manifest themselves during the life of the building. One recent example will illustrate this. A new sports/leisure centre had provision for separate changing facilities for the users of the swimming pools. The male changing room was provided with cubicles set in the middle of the room. These extended to about 1.3m above floor level. Each cubicle was just large enough for one person. Around the edge of the room were benches and clothes hooks. Doors supported by a pair of hinges allowed access into the cubicle. Within one month of the complex opening, over 60% of the cubicle doors were either completely missing, having fallen off, or were badly damaged. This was not due to vandalism but to the normal wear and tear that is associated with changing rooms. The problem was that the doors and cubicle sides were constructed in plastic laminate faced chipboard, thickness approximately 15mm. The screws for the hinges were totally inadequate to resist the swinging of the doors in reasonably heavy use – they pulled out from the chipboard. Under pressure from high – spirited youngsters the doors could not cope. A facility to ensure privacy for those who wanted it had become virtually useless. The reason must lie in the choice of cubicle design and materials, which were clearly not suitable for their function. This leads on to problems of maintenance. Customers can now say that the sports complex is not properly maintained; indeed with broken doors in the changing rooms, it could be seen as a safety hazard. But the fault is not due in the first instance to inadequate maintenance (though the doors had not been replaced a year after the opening of the facilities) but to bad initial design. Even if the doors were replaced it is inevitable that they would soon fail again. This is a clear case of a design decision directly affecting aspects of maintenance.

It is worth noting here the guide to good practice published by the CIOB's Maintenance Practice Committee, *A Guide to Maintenance Management*. This contains the following recommendations:

Design/maintenance relationships
New buildings are the product of the client, design team and the construction team.

It is essential during design and construction that all three parties understand the requirements of the proposed building. At the design stage decisions are taken which will affect the future use of the building.

Decisions affecting the future all-in costs of a new building are influenced by actions on the part of (a) the client and his staff/design team (b) contractor (or type of contract) (c) maintenance manager.

Action by the client and his staff
The client should:

(a) When selecting his architectural/design team for the new building bear in mind the maintenance performance they have designed previously.

(b) Supply the architect with a full brief of his requirements for the new building, following consultation with his own maintenance manager.

(c) Set up a small team from his staff to develop his ideas and requirements and to co-operate with the design team during the planning stages.

(d) Allow the design team sufficient time to consider all his requirements and prepare well reasoned schemes.

(e) State his policy towards maintenance (capital/running cost relationship) and provide the architect with details of any scheme of component standardisation, particularly electrical and mechanical equipment which operate in existing buildings.

Action by the Architect/Design team
The design team should:

(a) When studying the brief take account of the client's maintenance policy and standardisation requirements.

(b) When no detailed brief is provided, obtain from the client's staff the necessary basic information to enable them to proceed with a composite solution.

(c) Design the building to meet the client's functional needs, with due regard to the future all-in costs.

(d) Aim to complete the working details before the specification and bill of quantities are prepared and ensure that the responsible project architect checks the drawings for good construction, particularly in such elements as weatherings, expansion joints, throatings, laps, overlaps etc. and signs them.

(e) Be selective in its choice of materials for external use and their correct jointing.

(f) Endeavour to see that when new, untried materials are to be used they will not fail in position and that any special manufacturer's instructions or recommendations on fixing are known to the contractor.

There is as much responsibility for the future of a building on the client as on the architect/designer. It is the interrelationship between these two that will determine the levels of future maintenance. In the case of the unaware client it falls upon the architect to advise upon the long term effects of initial technological decisions. The above check lists should provide adequate prompts to designers.

The recommendations put forward to help the client are the key to any effective maintenance effort. If the client only pays lip service to the idea of maintenance reduction, then the architectural building team are less likely to produce technological solutions which are influenced by long-term considerations. If the client uses the legitimate excuse that the budget does not run to taking such a long-term view then the architect and builder should make clear to the client the possible maintenance repercussions. In the best situation, points (c), (d) and (e) under 'Action by the client and his staff', are the ones which will have a significant effect upon the design. Obviously, the client may be unable to involve a team of people, but in many cases one person will be enough if he is well briefed, knowledgeable and can liaise effectively with the design team. This must be done at the earliest stages to ensure full integration of the maintenance elements.

A major problem experienced by architects is the relative lack of time allowed by clients for the preparation of drawings and the final selection of materials and technologies. In the course of the design stage much time is spent on the feasibility of alternative schemes and in then going to the local authority for statutory approvals. The time between approval and when the client wants the job to start is least as possible. If the basis for technological decision-making is not firmly grounded then, in the haste to meet deadlines principles may be set aside. It has been reported that on one large contract the client made such strict demands for the production of the full scheme drawings that the principal architect had to employ 100 extra architects to meet the time limits. Will all these people be aware of the client's needs with regard to the policy of maintenance, or any other criteria for that matter? It may be said that clients who make impossible demands deserve to get an inferior building, but construction professionals do have a duty to tell the client the possible repercussions of his actions upon the building. After all, building is not just for the benefit of a particular client; the majority of buildings are sited in public areas and society as a whole has to enjoy, or otherwise, their existence.

Beddington ('The architect and building maintenance. The present position and a look ahead' by N. Beddington, in *Building Maintenance – Present and Future*, CIOB Maintenance Information Service No. 1, 1977) has identified six areas where design can materially affect aspects of maintenance. These are:

(a) *Materials*

Good maintenance is encouraged by a high standard of materials and design; an understanding of materials and appreciation of their behaviour in use is therefore essential. Each building type is governed by different considerations, but always the appreciation and study of cleaning and maintenance processes should be stressed. Also, manufacturers should be consulted for recommendations where these are known. External cleaning and repair, window cleaning and roof maintenance, all need consideration in the design stage.

(b) *Thermal insulation*

The increased thermal insulation required by the Building Regulations, higher temperatures, the popular demand for central heating and the disappearance of open fireplaces, plus the use of modern plastering techniques, have resulted in non-porous structures and condensation forming in rooms. Thus careful attention to new construction methods, e.g. vapour barriers and ventilation, is essential.

(c) *Design detailing*

Much has been made of defects in design detailing and the problems of rectifying them during construction (BRE Current Paper, 'Quality in buildings'). Architects must be reminded continuously of the need for attention to weatherings, throating and expansion joints, tolerances and correct jointing between different materials to allow for movement; there is unfortunately a tendency towards blind faith in the properties of mastic, rather than in traditional jointing details.

(d) *Vandalism*

Sturdiness in construction is the best precaution against damage. Badly lit and poorly finished communal areas attract the vandal. A higher standard of finish and amenity tends to be respected, as industry has found in its buildings.

(e) *Cleaning*

Before selecting materials it is worth preparing a typical check list, and considering the cleaning equipment, use, storage, service outlet positions etc. and the variety of cleaning operations and materials which may be required. Too many different materials needing different cleaning processes are a continuing nuisance to the occupier. Standardisation of sanitary fittings, ironmongery, light fittings, lamps, tubes etc. will facilitate replacement and the holding of stocks. To go for the cheapest may well not be an economy; ironmongery is a good example, and the amount of wear fittings are likely to experience must be taken into account when making a selection.

(f) *Services*

It is of the utmost importance that mechanical services engineers should consider the question of maintenance in the same way as architects do. Convenient access to servicing, consideration of materials in relation to their life, access for easy removal and concise routing of the services, positioning of the access panels, all generally leave much to be desired. Mechanical services engineers have a great deal to offer if they work seriously to observe these points, and discuss and agree them with architects. With services representing an increasing proportion of the building contract, upwards of 50% in some cases, the convenience and comfort of the building in use will depend more and more on planned and adequate services.

In the conclusions to her paper Beddington says that there still seems to be a lack of attention to building maintenance. The reasons for this are complex and include a contradictory fiscal system of taxation which needs reviewing, existing methods of financing and erecting buildings, types of ownership and occupation, as well as its lack of perceived significance to the client and architect. It is not yet accepted that a properly administered and allocated maintenance budget will arrest deterioration and subsequent high repairs or replacement costs, as may higher initial expenditure on fittings and materials. The architect should have maximum feedback and become more involved with maintenance service for the buildings in use. This will give a new dimension to the thinking of the architect not already concerned with such things, and architectural education should take this into consideration.

Another paper (Maintenance Information Service paper, Autumn 1977 by D.J. Cooper – 'Maintenance management today and tomorrow') showed how links were made between designers and maintenance in the Greater London Council. (It is to be hoped that initiatives like this will not be lost with the dissolution of the GLC.) The maintenance branch of the GLC was represented on the Development and Materials Committee, which sat under the chairmanship of the Technical Policy Architect. This committee consisted of the senior representatives from the structural building regulations, quantity surveying and construction divisions of the architect's department, with similar senior officers of the research, development and materials groups, the scientific branch and the maintenance branch. The committee was concerned with the development of new methods of construction and the assessment and performance of new and existing components and materials. The findings and recommendations of the committee were published in the *Development and Materials Bulletin*.

The branch was also represented on the following:

(a) Standards Review Panel. The panel considered such matters as minimum room area, provision of all types of fittings, and heating and lighting requirements.

(b) Window Working Party. Because of the high volume of replacements (£2m per annum at 1977 prices) due to wet rot and poor design there were continuing discussions concerning types and detailing in an effort to limit the problem.

(c) Contract Documents Panel. This was responsible for decisions leading to the revision in 1977 of the standard contract preambles.

It is obvious from the above that the GLC considered that the link between design and maintenance is not just an ephemeral thread. As a large organisation taking responsibility for client, design and maintenance functions it could create a structure to ensure adequate feedback and adoption of experience.

A major problem encountered by designers is the acquisition of meaningful information about the maintenance requirements of their buildings. Generally, the architect has little to do with the building after the defects liability period has elapsed. The client takes over its management and will only recall the architect in the event of a major difficulty. After the passage of years the architect's involvement is likely to become less. The architect will rarely be able to monitor the building in use. In the resulting absence of data it is not surprising that the architect could make the same errors in technology many times over. Skinner and Kroll have done some interesting work in collecting data on maintenance ('Maintenance feedback', N.P. Skinner and M.E. Kroll, CIOB Maintenance Information Service no. 18, 1981). They managed to produce upwards of 1500 different job identifications under 100 main element headings. Data was collected from a number of housing estates over a number of years, and was recorded and analysed with the use of a computer. From this it was possible to extract those items most frequently occurring; the most costly ones; the ones re-occurring, and so on. The authors consider that the potential benefits arising from collection of such data are:

(a) it helps to improve existing maintenance procedures;
(b) to improve the design of new buildings.

If studies such as these were undertaken as a matter of course by building owners with large numbers of properties a useful body of knowledge could be generated.

SUMMARY

Design decisions made regarding the construction of the building have a major influence over its performance. Giving some attention to maintenance aspects will undoubtedly produce benefits for the efficient functioning of the building and its services. A positive approach to recognising the long-term effects should be made by both client and architect. Where the client is unaware of the effects of initial design decisions on maintenance, the architect (or builder) should be prepared to explain. In order to create good designs, feedback from buildings and details of their maintenance histories are needed. The effective dissemination of this information is vital to ensure mistakes are not repeated.

QUESTIONS

1. Describe an organisation structure involving client, architect and builder which would ensure maintenance technology is considered at the design stage.
2. Identify a technological detail within your experience which has incurred excessive maintenance and show how a better design could have eliminated this.
3. What responsibility has the builder for advising client and/or architect on the maintenance aspects of proposed designs?

4. Buildings as Waste Makers

Harper (*Building: The Process and the Product*, D.R. Harper, 1978) has formulated the concept of buildings as waste makers. He sees them as being in many cases extremely wasteful in their use of energy. There are two main kinds of extravagance:

(a) that induced by the inflexibility of systems within the building;
(b) careless use of energy by the building's users.

In the former kind there is the situation where the building design – large open areas, high floor to ceiling dimensions – denies the efficient use of energy, or the heating and lighting systems cannot be adequately controlled to meet local service needs. Buildings designed and built under different social and economic priorities will reflect prevailing views on the use of energy. In the past the demand for space and the 'right' architectural proportions created large rooms in area and height. Into these were placed massive systems of space heating, such as large bore pipes carrying steam. With the relative abundance and cheapness of coal, the cost of running the boilers was of secondary importance. In some instances this lavish attitude to the use of energy is still prevalent, even since the 1970s oil crisis which led to a reappraisal of energy resources. There are many instances where lighting can be seen to be excessive. In addition the general public as energy users are usually careless in the control of energy, even when the switch can be operated directly.

In the UK there are about 19 million dwellings, of which approximately half were built prior to the 1939–45 war. These homes were built to standards which did not recognise the need to conserve energy to the degree now prevailing. Despite the inducement of grants to upgrade thermal insulation to these properties, the majority still do not meet present-day standards. They can be said, therefore, to be waste makers. Without extensive work involving forms of insulation to the buildings' external walls, conversion to more efficient and controllable heat sources and a general sealing of cracks and edges around openings, the loss here will continue. For large buildings, or a number of buildings under the responsibility of one organisation, it is possible to carry out energy audits. The CIBS Energy Code has already been outlined. The four volumes of this code include the process and formulae for the calculation of the energy demand of a building. Although the code is primarily designed for new buildings its principles can be applied to existing buildings. The determination of the energy demand of any building can be undertaken. This can then be compared to the existing standards. If there is a wide discrepancy between the two, that is, if demand exceeds recommendations, then some remedial work ought to be carried out to prevent the waste of energy. It may be useful to carry out this appraisal every few years. If there is a deterioration in the demand levels, i.e. demand is increasing, the reasons for this should be investigated. It could be that the service plant for this should be investigated. It could be that the service plant is becoming inefficient and is requiring a greater energy input to sustain the required temperature levels. It may be that the building fabric is deteriorating, allowing weather penetration which requires energy to combat its effects. Or it may be that the building's users are becoming careless or profligate in their use of energy. Whatever the reason, an energy calculation will give the building owner (or payer of energy costs) an indication of how that building is performing with regard to its use of energy.

CLEANING

A necessary performance requirement in any building is to keep it clean. A domestic house will differ in its demands for cleanliness as compared to a suite of offices. For those people who carry out the task it is one that is never-ending. In some environments it is even essential to clean the air, for example hospital operating theatres and office computer rooms.

Those buildings with a great variety of floor or wall surfaces are very wasteful in cleaning terms; they require a large number of different methods, such as brush, mechanical polisher, wet mop, and vacuum cleaner, with their complementary cleaning agents. Problems arise at the junctions between different finishes – how are these to be maintained? There are dangers to the building's users in that they may carry one cleaning agent on the soles of their shoes on to another surface which may cause some slipperiness.

Those surfaces which need constant cleaning in order to retain their freedom from slipperiness can be seen as wasteful in energy and materials. Poor quality carpeting can wear rapidly, causing a danger to walkers and requiring frequent renewal.

Although cleaning using mechanical aids can be much quicker and more effective the use of energy is higher. Most will require an electrical power source.

In the design of buildings the implications of cleaning for the overall use of energy need to be considered, and its contribution to the unnecessary production of waste.

SERVICING

To be efficient, effective and safe most plant and equipment in a building requires regular servicing. The frequency of this servicing can be considered wasteful if:

(a) modern equivalents require considerably less service visits;
(b) breakdowns occur between service visits;
(c) it requires the recurrent renewal of parts.

How often is 'frequent' can only be judged against a present-day equivalent piece of plant undertaking the same function. For example, a modern lift might require just two visits a year. An old lift might require five or six visits, a much higher frequency. This has to be weighed between the service visits it could be:

(a) that the parts are unsuitable for their tasks;
(b) that the service visits are not based on preventive maintenance policies.

In both cases there is cause for concern in that they are contributing to wasteful practices. The constant renewal of parts in itself is wasteful.

RUNNING

Energy is required to run the majority of services provided in a building. Power is required to move lifts, provide heating, give lighting. The over-provision of these facilities can lead to unnecessary waste. Where servicing is not adequately carried out, an undue increase in energy usage will result; this is due to the plant and machinery not running efficiently. For example, a blocked gas jet to a boiler could cause insufficient combustion leading to wasted fuel. The boiler also has to work harder in order to maintain the required levels of heat.

SUMMARY

As buildings are a major user of the nation's energy output they ought to be viewed as potential waste makers. This being the case then it can be readily appreciated that there is a need for well thought out strategies for cleaning, servicing and running a building, based on a quantitative analysis of the energy used. This analysis can highlight areas of excessive demand, and remedial action can then be devised and implemented.

QUESTIONS

1. Should an industrialised nation encourage better levels of energy conservation for existing buildings, rather than attempting to develop and produce cheaper sources of energy?
2. Discuss at what point it might be better to replace an inefficient service item rather than service it continuously.
3. Taking the approach that buildings are waste makers discuss how this might affect the technology of service installations.

5. Maintenance of the Fabric

INTRODUCTION

In this chapter two concepts will be considered which will have some effect on the maintenance of a building's fabric if they are used during the design and build stages. Even if they were not adopted initially they can provide a useful check when looking to create strategies for aftercare, as described in the next chapter. Here the effect of terotechnology and life cycle costing on maintenance will be discussed.

TEROTECHNOLOGY

Terotechnology is defined as 'a combination of management, financial, engineering and other skills applied to physical assets in pursuit of economic life cycle costs'.

A major study, based on the principles of terotechnology criteria, was carried out by the Local Authorities Management Services and Computer Committee (LAMSAC), together with the Department of Industry, and published as a report in 1981, *Terotechnology and the Maintenance of Local Authority Buildings*. This study was a fact-finding exercise aimed at acquiring more knowledge than was previously available about local authorities' existing practices and procedures for controlling aspects of the maintenance process. The terms of reference were to 'undertake a detailed survey of the methods used by local authorities for maintenance expenditure, and the effects of making the expenditure and the financial controls used'. It was restricted to the non-housing field.

There were over forty conclusions, commenting on such aspects as organisation and responsibility; staffing; property inspections; property records, budgets; priorities; tenders; contracts; accounting; analysis of expenditure; planned work and feedback. This last is worth quoting in full, as it related to the points made in the previous chapter on design. 'Lip service is paid to the necessity for feedback from maintenance to design, but very few effective systems exist for this purpose. Maintenance personnel in general despair of getting designers to learn from past errors, at least by purely persuasive methods.'

The concern is how an approach based on terotechnology criteria can aid maintenance procedures. Some of the recommendations presented in the report are presented below. This will enable the range of considerations within this concept to be appreciated.

Centralised control

There is a need for centralised control over local authority maintenance functions. Full communication and co-ordination is only likely to be achieved where all personnel engaged full-time on maintenance work, from whatever discipline, are within the same section.

Staffing

Urgent investigation is required into appropriate levels of staffing in relation to maintenance workloads, with particular regard to the balance of duties and functions as between technical and support staff.

Property inspections

Systems for carrying out and recording results of property inspections are in general capable of considerable improvement.

Maintenance budgets

Annual budgets should be formulated by reference to intended programmes of planned work, plus an amount deemed sufficient (on the basis of past experience) to cover likely spending on day-to-day and emergency items.

Tendering and contractual arrangements

Requirements for excessive numbers of tenders, even for comparatively small works, are time-consuming and costly. In appropriate circumstances, greater use might be made of multi-property and term maintenance contracts, producing considerable savings in administration.

Direct labour

Policies in regard to the use of direct labour for maintenance work should, in many cases, be carefully examined. Whilst DLO's have undoubted advantages the means whereby work is allocated to them do not always appear to be necessarily in the best interests of the authority, nor are they invariably conducive to efficient maintenance management and control.

Computerisation

Maintenance management would benefit from the development of computer applications in the following areas:

building records;
commitment and payment accounting;
specifications;
analysis of expenditure.

What can be drawn from this report? There are two main themes: one concerning the science of terotechnology as applied to maintenance; the other the implications of its recommendations.

The science

Terotechnology is very wide-ranging and is essentially concerned with the structure and systems which affect practice of the maintenance function. It considers the skills of the managers; whether or not they can be aided by using modern information technology; how the process of allocating work can be improved.

Implications

As the report notes, there is still little feedback from maintenance to design. If this does not happen in organisations which can, in theory, co-ordinate departments (as seen in the description of the GLC structure) what hope is there for effective feedback in sectors where a number of separate groups contribute to the construction process? In local authorities there is the opportunity to inform the designers of maintenance problems, but, according to this report, the designers tend to ignore them. In the case of a private client commissioning a building, the architect is not usually involved after completion. Only in rare instances will he be involved with aspects of maintenance. Here there is little opportunity for the feedback of information. But, if there was, would we find the same reluctance to consider maintenance in the initial design stages?

The report paints a sorry picture of the development of building technology. One way of learning is from mistakes and to ensure that they are not repeated. If there is little or no feedback then building technology suffers drastically.

LIFE CYCLE COSTING

This technique has been discussed in Part Two on Design. Here the relationship between life cycle costing and maintenance will be considered.

One of the uncertain areas within the formulae for calculating future costs is the estimation of the amount of maintenance required to a building or part thereof. One needs to be able to predict, with some confidence, the amount of maintenance so that alternatives can be fairly compared. This information can be obtained in the following ways:

(a) manufacturers'/suppliers' guides to maintenance cycles on their products;
(b) analysis of previous records of maintenance to similar items;
(c) an intelligent estimate based on experience.

In the majority of calculations it is the last that is employed, with perhaps the first. Records of maintenance are not commonly available, so all designers will not have access to this information. Even where records are available they may be in such a form as makes analysis difficult; or they do not take into account new products and practices.

By using life cycle costing techniques in the initial design phase attention is brought to the maintenance aspects of the building technology. Even though there may not be hard evidence relating to maintenance cycles, frequencies and costs, it will cause the designer to give an estimate of these factors. Maintenance becomes an integral part of the design. Bampton ('Presenting the owner's case', E. Bampton in *Building Better Buildings. Maintenance at the Design Stage*, CIOB Maintenance Information Service No. 3, Spring 1978) recommends the following procedure in order to reduce maintenance

The design team prepares a draft or several draft schemes with rough cost appraisals.

The design team prepares the final scheme with some alternative capital costs and the first cost-in-use assessment. (The client may also require a total life cycle cost appraisal to be made.)

Here we have a client advocating the use of life cycle costing to

3. PLANNED PREVENTIVE MAINTENANCE

This strategy is preferred by many as an overall policy to adopt in the pursuit of effective maintenance. In essence, it is maintenance which is carried out on a regular basis in order to prevent breakdowns and failures. It is allied to cyclical maintenance, as the fabric or services are not allowed to deteriorate to a point where they fail. Bushell ('Preventing the problem – a new look at building planned preventive maintenance', R.J. Bushell, CIOB Maintenance Information Service No. 11, Winter 1979/80) has given the following definitions:

Maintenance. A combination of any actions carried out to retain an item in, or restore it to, an acceptable condition. Note: the actions referred to are those associated with initiation, organisation and implementation.

Maintenance policy. A strategy within which decisions on maintenance are taken.

Planned maintenance. Maintenance organised and carried out with forethought, control and the use of records to a predetermined plan. Note: preventive maintenance is normally planned. Corrective maintenance may or may not be planned.

Preventive maintenance. Maintenance carried out at predetermined intervals or to other prescribed criteria and intended to reduce the likelihood of an item not meeting an acceptable condition.

Running maintenance. Maintenance which can be carried out whilst an item is in service.

Bushell goes on to describe a procedure for the implementation of planned preventive maintenance. In the first place the client must have a policy which is initially broad enough to include a commitment to planned maintenance. In its detailed content it should advocate prevention rather than cure, providing the following criteria are met:

- it is cost effective;
- it is wanted to meet statutory or other legal requirements;
- it meets a client's need from an operating point of view;
- it will reduce the incidence of running maintenance necessitating requisitions for work from the user;
- there is a predominant incidence of work from the craftsman rather than pure inspection.

There are four stages in the development of a planned preventive maintenance programme:

(a) Analysis of requisitions

If an analysis of the previous 12 months' requisitions is undertaken it will give a good indication of where there is a recurring problem. (See 'Maintenance feedback', N.P. Skinner and M.E. Kroll, CIOB Maintenance Information Service No. 18, 1981.) Also an analysis of accidents will indicate danger areas and may also be cost-effective when balancing the cost of meeting accident claims, investigations etc. against the cost of maintaining an element of construction in a preventive programme.

(b) Survey of the estate to establish needs

This should be carried out on a regular basis to establish the overall requirements. The survey should identify and quantify the work-load, and also give some detailed information on certain elements of the building, for example: roofs, rainwater goods, drainage, main structure, windows and doors, floors, internal plumbing, decoration, car parks, signs etc.

(c) Programming

The programme is drawn up from the data provided by the analysis and the survey. It is geared to the frequency of the tasks and the labour and finance available. The greatest problem lies in assessing task frequency, which usually ends as a compromise between the money available and the risk attached to inadequate frequency. Consequently, priorities must be assessed carefully.

(d) Implementation

There appears to be a general lack of commitment to introducing a programme of preventive maintenance, but its advantages are:

(i) the client can see what service he is getting and what risks are protected;
(ii) work is organised and therefore controlled;
(iii) requisitions and the running of maintenance on a crisis management basis are reduced;
(iv) statutory legal and professional responsibilities are met;
(v) direct or contract labour can be used to execute the work;
(vi) the programme is flexible, subject to review and can be implemented in stages;
(vii) essential estate management records are obtained;
(viii) facts become available to compile a replacement programme.

Lee has identified two complementary and interacting systems operating within a planned programme (*Building Maintenance Management*, R. Lee, 1976). These are called 'schedule' and 'contingency' systems (see Fig. 4.1 for a simple diagram showing their interrelationship). The *schedule* system covers items which tend to deteriorate at a generally uniform rate and do not have a high degree of urgency. It can be implemented in three ways:

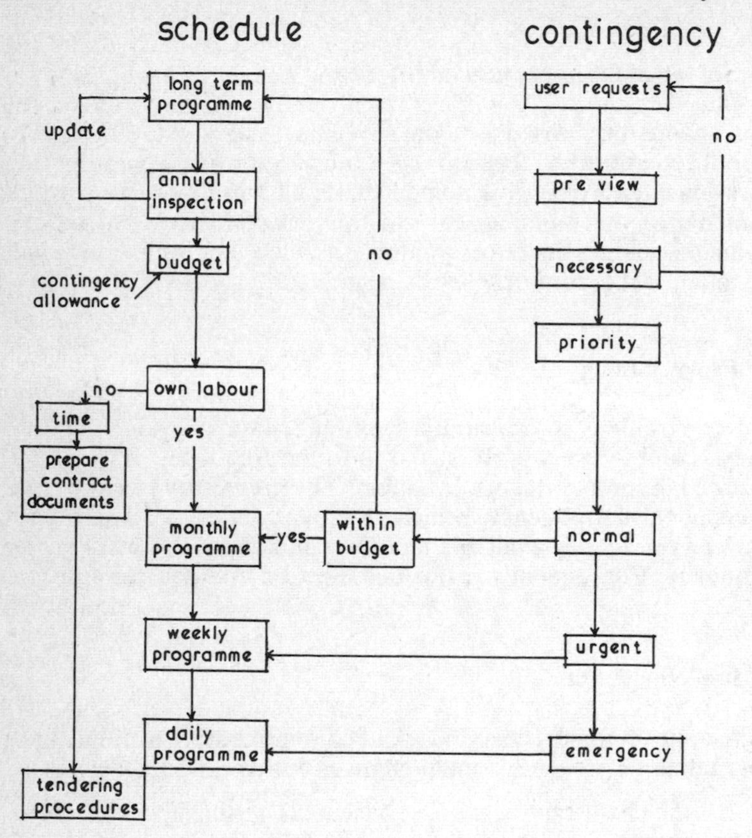

4.1 Maintenance strategies (from Lee, R., Building Maintenance Management, *Granada, 1981)*

(i) by scheduling the work to be carried out at predetermined times. This applies where the incidence of failure can be predicted with some accuracy, or where periods are fixed by statute or contract – for example the terms of a lease requiring redecoration at fixed intervals;

(ii) by scheduling inspections to be carried out at predetermined times to detect failures or the imminence of failures. It could be that the exact time of failure is not known and, therefore, an inspection will determine whether or not the work is necessary;

(iii) by scheduling work and inspections to be carried out at predetermined times. This applies where a confident prediction of necessary work can be made, but an inspection is necessary to find the extent of any further work.

The *contingency* system is based on waiting until a complaint is received from the user before any action is taken. If the frequencies and types of complaint are analysed, procedures for dealing with remedial work can be planned even though the timing is uncertain. A necessary part of the system is the need to introduce a delay period between the receipt of the request and the actual execution of the work. This will permit the grouping of similar items to regulate the flow of work to the workforce. The main difference between schedule and contingency maintenance is that the time between modification and execution is greater for scheduling maintenance. The choice between these two systems can centre on the following factors:

(i) the predictability of failure. Burst pipes can only be dealt with under a contingency system. Components which deteriorate at a known rate can be scheduled for repair or inspection before they actually fail.

(ii) the reporting delay time. The interval between notice by a qualified inspector and the time when the occupier would report the defects. It depends mainly on the degree of inconvenience which the defect causes to the occupier, not the measure of the seriousness of the defect. If the reporting delay is less than the economic period for carrying out inspections the work must by necessity be dealt with on a contingency basis.

(iii) the rate of deterioration of the component and the corresponding increase in the cost of rectification. Also to be considered is whether it is the type of failure which is likely to prompt an early response from the user.

(iv) the extent to which the user can be relied upon to report significant defects. This will depend upon the nature of the occupancy and the attitude of the user to the condition of the building.

In any system of preventive maintenance there will be some delay between receipt of the user request and the execution of the work. These delays could be determined by any of the following:

(i) safety considerations;
(ii) user satisfaction. The contingency system does allow the users to participate in the running of the building;

(iii) effect of failure on the primary activities of the organisation. Where, say, a factory fails it may affect production processes and cause heavy monetary losses. Preventive scheduled maintenance would then be the system operated;

(iv) dispersion of job activities. How close together are the maintenance tasks? Travelling times can be as much as 40% of the total job costs;

(v) cost growth. Generally, the longer a defect is left unattended the more expensive the remedial work;

(vi) misuse of property. A poorly maintained building could encourage misuse and vandalism, thereby necessitating further remedial work.

In this review of planned preventive maintenance two approaches to the exercise of identifying what is to be done, and when it is to be done, have been considered. The first sets out a series of steps for the implementation of a planned programme. The second recognises the need for a planned programme (scheduled system) but also sees it necessarily interacting with a contingency system.

4. UNPLANNED EMERGENCY OR REQUISITIONED MAINTENANCE

This is maintenance which is left until there is a breakdown, reported by client or maintenance staff. With some items it may be expensive owing to loss of facility or damage to secondary items. It is also expensive in non-productive time for the maintenance craftsmen through increased travelling, and in diversion of resources interrupting the maintenance programme. It would be too costly to obviate this type of maintenance altogether, and an economic balance must be sought.

5. MINOR NEW WORKS

This is the last type of broad aftercare strategy. It includes small alterations, additions and upgradings required by the client. In effect this can often form a considerable part of the total programme for aftercare, at the expense of funds and resources for basic maintenance. Every effort should be made to ensure an economic balance between minor works and pure maintenance. Ideally, every item under this heading should be justified by a cost benefit analysis, as money can be wasted on unjustifiable jobs.

SAFETY IN MAINTENANCE

An essential part of the process of monitoring maintenance should be the periodic review of safety. This should include:

(a) the safety of workers carrying out maintenance;
(b) the safety of users and visitors to the properties being maintained;
(c) the safety of the building structure and its equipment and services.

With the growth in safety legislation, the importance and amount and cost of maintenance has increased. The Health and Safety at Work Act, the Occupiers Act and the Defective Premises Act are relevant to this aspect of monitoring.

Bushell ('Assessing maintenance priorities. Guidelines based on Health Service experience', CIOB Maintenance Information Service No. 17, 1981) lists the following priority factors for an effective work programme:

Safety
Essential service
Statutory requirements
Security
Initial cost
Revenue saving
Spares availability
Alternative source of supply
Delivery time
Manpower
Public relations.

As may be seen, safety is top of the list. The reasons given for ranking it at number one are that it is a prime duty to provide a safe and healthy environment, and that potentially dangerous or hazardous items must be a first charge on resources available. Structurally dangerous items should not appear on any maintenance list as they must be dealt with as soon as they arise.

From this viewpoint maintenance is seen as an activity which has as its main concern the safety of the users of the building.

CONTROLLING MAINTENANCE

Whatever aftercare strategy is used there must be some form of control. This can be exercised in two areas: finance and progress.

Financial control

In the first instance there must be a budget. This is prepared either on the basis of the previous year's expenditure, or, where figures for this do not exist, on an estimate of the maintenance costs for the coming year. The accuracy of the estimate will depend on the depth of information and data available for:

(i) nature and extent of work;
(ii) conditions under which the work will be executed;
(iii) mode of execution;
(iv) costs of employing labour;
(v) prices of materials.

Long term estimates can be devised using financial criteria – expressing maintenance costs as a percentage of:

(i) construction costs;
(ii) production costs;
(iii) occupation costs;
(iv) profitability;
(v) costs per unit of accommodation;
(vi) costs per unit of floor area m square;
(vii) costs per unit of building volume m cube;
(viii) costs per building element;
(ix) costs per functional system.

Short term estimates can be prepared using:
(i) analysis, breakdown of activity into labour materials and plants;
(ii) judgement, based on experience of doing similar activities;
(iii) slotting, classifying jobs within time brackets and adding a percentage for materials.

It is recommended that the budgetary period should be related to the rate of deterioration of a significant element of the building. In the case of external painting, for example, a five-year budget period should be used. The annual programme can then be seen as part of a continuing series of work matched to the organisation's cash flow.

Where a budget has been prepared it would be possible to monitor the actual costs against those forecast. Ideally, individual major items should be estimated and allowances made for contingency items.

Controlling progress

The control of progress can take two forms:

(a) progress of requisitions from the client. This system should ensure that requisitions are registered, and that the work required is inspected or assessed, then authorised. Arrangements to be made for access and then the work ordered.

(b) control of work ordered. The system for the control of work should ensure that there is a record of each job or project and of the orders placed on each, against which can be shown the progress achieved at stated intervals, say monthly. A system of work inspection and quality control should run in parallel with a system of work control. Reasons for any delays should be investigated and remedial action taken if necessary.

COMPUTERS AND CONTROL

Large organisations concerned with maintenance, such as health authorities and industrial complexes, are finding computers a very useful aid. Pettitt ('Computer aids to housing maintenance management', R. Pettitt, CIOB Maintenance Information Service No. 10, 1979) describes a system used by a development corporation. Its objectives are to:

(a) record, store and process every item of repair. A detailed financial breakdown and analysis can then be made;

(b) analyse the frequency of repairs. This will indicate future patterns of repairs.

Skinner and Kroll, in the paper already mentioned, put forward a coding system which can closely identify the items, components and elements of a building. In this way a detailed picture can be drawn up of those items and the manner in which they are failing or requiring maintenance. This will give a sure indication of the technological problems recorded in the maintenance histories.

Some health authorities now operate their maintenance activities wholly on computer programs. These will record and monitor work as it is done, giving costs, frequencies etc. and also priorities for activities. The planned preventive items are stored in the program. For each work period the items are called from the computer. The list is in a ranking order of priorities. Job tickets are issued directly from the computer. If, owing to a heavy workload, staff absences or emergency work, some items are not done, the following week the computer will again list these (together with the new items) and give them a high ranking. Once an item has been recorded as complete it will remain dormant until the next regular time period. The system is also able to give target times for the execution of the work, for bonus purposes.

With the use of computers for feedback to design, and in the everyday control of maintenance activities, one can be more confident of tighter control. They are useful in quickly highlighting problem items and giving an indication of the types of problems occurring. Obviously, this presupposes the correct input of information into the program!

SUMMARY

In this chapter, on strategies for the aftercare of buildings, five broad approaches were used, namely cyclical, non-cyclical, planned preventive, unplanned requisitioned and finally minor new work. The planned preventive maintenance strategy was highlighted as this system promises the best long-term advantages to the client. Also considered was the importance of safety within maintenance (in one view the overriding reason for maintenance activity). Additionally, some ways in which we can control and monitor maintenance functions, and in which the computer can aid these processes, were introduced.

QUESTIONS

1. Discuss the implications for technologies employed by a building tenant's organisation if it can be sure of its workload on a year-to-year basis.
2. Give reasons why maintenance work might be done only as a last resort.
3. Discuss how minor works activities could cause the failure of important building elements.
4. Comment on the need for all maintenance to be carried out via a planned preventive programme. Is this too idealistic?

7. The Concept of Spare Parts

In Parts Two and Three on Design and Production the idea was introduced that building is tending towards an assembly process rather than in situ construction with basic materials. It has been seen that components have an increasing role as constituent parts of a structure. These components are manufactured off site, delivered and fitted together in place. They can be bought off the shelf, with some being available from a number of manufacturers. The large DIY supermarkets stock a full range, from windows to sanitary goods, to central heating systems, to door locks.

When one considers the maintenance aspects of a building assembled and built with components it is easy to envisage the relatively simple replacement of worn out parts. In theory a component building should lend itself to this. In practice, at present, there is not much difference in the technologies used. Where in traditionally built structures there are walls, floors, roofs, windows, doors, services etc. these will be similar in component structures. Where a window was built into a brick wall as the work proceeded, and in effect became an integral part of the wall, so too will a window incorporated into, say, a steel frame structure with a lightweight cladding. The fixing of the window into the cladding and its framing will be just as secure as that of the brick wall window. Indeed, there are examples where component construction could hinder the replacement of items. The present day use of chipboard sheet flooring is not conducive to easy replacement, either in part or in whole. In the case of local damage to the floor the whole sheet will need to be replaced. Also, if there are problems with the services which are under the flooring, then access will be difficult, whereas the traditional tongue and grooved flooring boards did allow piecemeal lifting and replacement.

Where the concept of spare parts is likely to be fully realised is in the provision of the services to a building. To take a simple example, the waste pipes to sanitary fittings can be easily replaced. Central heating boilers can be completely replaced, or just internal parts. Obsolete lifts can be taken out and replaced with up to date units.

It is also possible to envisage the external fabric of a building being replaced. The fabric would need to be non-loadbearing and independent of the structural system. A present example is of profiled metal sheeting. Its fixing to a subframe is by common nuts and bolts; therefore, if it requires replacement, the whole panel can be removed easily. A single panel can be replaced or the whole of the building envelope.

There are now a number of large buildings which were designed by architects with the deliberate intention of exposing all the major service functions; examples are the Lloyds underwriters' building in London and the Hong Kong and Shanghai Bank in Hong Kong. In both these buildings the toilet facilities are self-contained room-sized modules; they were bolted to the main structure as units consisting of walls, floor and roof, with all fittings in place. They were then connected to the main structure and service runs. If they become obsolete or suffer a major failure they can be replaced as units. This is not expected, but provision of the possibility is there. With the Hong Kong Bank the manufacturers of the major components were asked to produce extra items; these will be stored and used to replace any that fail on the building. In other words, 'spare parts' are available.

The exposure of service pipes on the outside of the building allows renewal and maintenance to be easily carried out. Where the services are contained in a building, usually in an enclosed duct, there are undoubtedly problems regarding access. Where, however, the main functions are on the outside of the building, albeit insulated and protected with special coverings against the elements and to give a pleasing appearance, wholesale replacement or modification is possible, as is all round access for servicing and maintenance. To see large pipes running vertically and horizontally around the building may not be to everyone's taste, but from a functional point of view there are advantages, and the concept of 'spare parts' technology may thereby be realised.

SUMMARY

Examples exist of buildings assembled with components which, in theory, may be removed and replaced with spare parts. In practice this may be difficult to achieve. Some large buildings with sophisticated services and their attendent carcassing left exposed will allow the concept of spare parts to operate. Whether or not it will be feasible actually to replace these items remains to be seen. It is possible from the technological point of view, but will it be economic? Will this concept be adopted for smaller-scale buildings?

QUESTIONS

1. If buildings are assembled using large-scale components, such as complete toilet facilities, discuss the effect this will have on the technology of their maintenance.
2. Discuss the implications of being able to have a 'store' of spare parts for the major elements of a building.

8. The Recycling of Buildings: Choice of Method

There are a number of terms used to describe the recycling of buildings. In itself recycling can be defined as 'renovation for a purpose other than its original use'. This implies the reuse of certain elements or components, with the foundations and structure left intact. The point where recycling ends and new build starts will be discussed after an attempt has been made to define some common categories of recycling.

1. REHABILITATION

This term is usually used for work carried out on dwellings. It covers work intended to make buildings fit for human habitation by present-day standards. This may include such items as reroofing, providing a higher degree of thermal insulation and updating sanitary facilities. In some cases it may mean the arrangement of accommodation. For example, a dwelling with no inside toilet or inadequate bathroom may have to sacrifice a bedroom or living room to house these facilities. There have been schemes where local authority-owned property has been bought by a private developer and rehabilitated to a high standard for resale or renting. In one block the number of units may alter, becoming greater or smaller (generally the latter) than the original. One reason for sale is often that properties are unsuitable for local authority tenancies and, to avoid demolition, the building may be upgraded to suit a different group of people.

Rehabilitation can be used as a term to describe work to commercial buildings, where the work carried out is similar in nature to that just described. Generally, though, the following term would be used in such cases.

2. ADAPTATION/REFURBISHMENT

These terms are used where a building has some substantial work carried out in order not only to update the service functions but also to provide relevant accommodation which may satisfy a different user function. For example, it has been known for a

multi-storey car park to be converted into flats. An office building with obsolete services, inadequate movement corridors and facilities (e.g. lifts) could be adapted to suit current needs, i.e. the demolition of solid internal walls to allow an open floor which can then be partitioned as required, with flexibility for a future change of layout. Some floors could be removed, replaced or even inserted, depending on the requirements.

Reburbishment work may be limited to the rearrangement of internal layouts by the insertion of demountable partitions. It could also mean the virtual demolition of the structure, leaving some external walls on their foundations.

3. PRESERVATION/CONSERVATION

It is difficult to differentiate between these two terms, but the following explanation may go some way towards further defining this category. The cleaning of a building's facade, for instance, constitutes preservation. No changes are made, some remedial work may be done, but the intention is to prevent decay, or to bring back the item to its original appearance and state.

Conservation would be seen as going some stages further. The original materials, details, shapes, colours etc. are considered as sacrosanct. Any replacements or repairs must exactly match the originals. This may even extend to the original services such as sanitary fittings and heating system. To carry out this work specialist skills may be required, over and above those expected of a maintenance craftsman. The use of old techniques, tools and materials will demand a degree of skill which is now not commonly found. A thatching repair would fall in this category, as would the repair of ornate ceiling bosses in fibrous plaster.

THE OVERLAPS

In any given project it could be demonstrated that all the above categories apply. Take, for example, the restoration of a large old house. Is the aim to update the existing services and accommodation requirements (rehabilitation)? Or to divide it into separate flats (adaptation)? Or to have the external stone facade cleaned and repaired (preservation/conservation)? Whilst one of the above categories can indicate the major purpose of the work, they are not mutually exclusive.

Another 'grey area' is where, owing to a preservation order, an old facade to a building has to be retained but the building itself cannot be readily adapted to meet current functional demands. A new structure may be built, with new floors and roof, together

with services. In other words all that remains of the old building is the external envelope. New windows may be fitted. Is this refurbishment or new building, or even conservation? On balance it might be viewed as new build, with a degree of conservation in the retention of an important element.

FACTORS DETERMINING CHOICE

The life span of a building can be seen from two viewpoints, those of the economic span and the structural span. In the first the life of the building is expressd in monetary terms – at what point in time is it uneconomic for it to continue in existence? The calculation of costs should include capital and investment, financing, operation/running, maintenance, repairs and replacements, renovation, salvage or disposal, depreciation and tax write-offs. In the second, the building's life span is the period during which it remains structurally sound and safe. Generally, this period will exceed the period during which the building is economic. Therefore, if decisions are made on monetary criteria, buildings will be adapted, altered, rehabilitated or demolished before, in functional technology terms, they have run their full life. In the determination of economic life span, techniques such as value analysis and life cycle costing should be used to ascertain the quantitative data upon which decisions can be made. The factors to be considered in such an analysis are:

(a) *Purpose and use of building*
Is it fully used? Is its purpose still viable? Will the use continue?

(b) *Appeal of location, appearance and effectiveness for its future occupants*
Has the area changed in planning terms? Is the building's appearance (architectural style) unacceptable? Will future occupants be attracted to it?

(c) *Comparison with competitive alternatives*
What other similar use buildings are in the locality? Does this compare with them? Is it more, or less, expensive to run?

(d) *The economic projections – present and future*
Is the building now economically viable? At what point in time will it become uneconomic? Will any losses be sustainable?

(e) *Potential extent of changes to make the building more suitable to prospective tenants*
Under this factor can be included technological problems such as: fire hazards, statutory and insurance requirements. Any

need for special fire proofing? Are the internal partitions load-bearing? Is the thermal insulation adequate or does it need upgrading? Is the layout of the building easy to change? What is the full extent of the work to maintain, renovate, adapt or upgrade? Will the structure allow the inclusion of modern service requirements and be able to provide a suitable internal environment?

(f) *Nature of the construction situation*

Are the necessary labour, materials and plant available to undertake the work? Are the necessary skills available? Can a suitable work methodology be devised to bring about the desired changes in the building? Can these changes meet statutory requirements? What is the state of the market for the type of work envisaged? Will the costs be inflated?

(g) *Nature of financial situation*

Where will the money come from? What are the interest rates? What are the calculations regarding amortization of the money? Are there grants available? What is the influence of taxation? Will insurance premiums be greater or less after the work?

(d) *Life span variables*

What will be the projected life span of the building after the works? Will the various elements and service functions have different life spans? How will this affect the potential future users?

ENERGY

The energy implications of change of use or demolition of a building must also be considered. The concern with energy consumption has already been discussed in relation to design and production, but in deciding whether to renovate or demolish, one should bear in mind that if the building is demolished the great majority of the energy used in its construction will be lost. To compound this, a new building will require more energy in the replacement of the main elements. It has been calculated that the energy lost in the demolition of the building can be as high as 1900 joules/sq. m.

Against this must be set the possible waste of energy in a building which cannot be sufficiently upgraded to meet present expectations or requirements.

When planning heating and ventilation systems, they ought to be designed in such a way that they are capable of being converted from one fuel source to another.

CONSTRUCTION CONSTRAINTS

In considering construction methods the following set of constraints attributable to work associated with existing buildings should be recognised.

(a) *Limited or difficult access*
Generally, buildings for refurbishment etc. are situated in urban areas. They can be adjacent to other properties. The roads and surrounding streets could be narrow. The site itself may have no room for the storage of material etc. except within the perimeter of the building.

(b) *Limited work area*
Within the building, work areas may be limited owing to the need to store or move materials, or to the removal of floors or the installation of services.

(c) *Limited headroom*
Conventional plant and machinery may not be able to operate owing to lack of working height. Movement and placement of large items and components may be restricted.

(d) *Limitations of floor loads*
Floors designed and built in the past may not be able to sustain the loads due to materials storage and movement. The safe bearing capacity of the floors will need to be ascertained.

(e) *Limitation of work period*
If the building is located in a built-up area the adjoining building users may object to the disruption of the enjoyment of their environment. Heavy site traffic may inconvenience their access. Work may not be allowed, in the case of commercial areas, between 9 a.m. and 5 p.m. In the housing area, work may be restricted to 8 a.m. to 6 p.m.

(f) *Restrictions concerning dirt and dust*
As adjoining properties might be affected, especial care should be taken to reduce dust and dirt pollution, for example by providing enclosed shutes to take waste from high to low levels for ultimate disposal.

(g) *Restrictions concerning noise*
Working in urban areas means that noise levels must be kept to the minimum. Statutory requirements will have to be met. There may also be problems of noise owing to working in confined spaces within the building; extra protection to the ears of the operatives might be required.

(h) *Restrictions concerning fire hazard*
When working in an existing building the risk of fire can be greater than in new work. Most of the building will remain standing throughout and its fire load could be high. Whilst the structure is exposed during upgrading the fire risk is at its highest.

THE TRANSIENCE OF BUILDINGS

Buildings do not have an inherent right to exist for ever. As society and technology changes so does the demand for buildings. It is likely that in a hundred years' time the great majority of buildings now standing will not exist. Up to that time many could have been renovated or refurbished once or a few times. It has been estimated that the average life span for one use of a commercial building in a city centre in the USA is 16 years. Decisions as to whether to maintain, adapt or demolish are centred on economic and functional life spans. There are cases where the economic and functional life spans of the building still have a considerable time to run but where the building may be demolished nonetheless. This may be due to:

(a) *Provision of new roads etc.*
In an industrialised nation there is a constant demand to upgrade the roads or provide new roads. Unfortunately, where these are required in urban areas there is an inevitable conflict between road widening/provision and the retention of buildings. Government, who generally sponsor major road works, has to plan the overall development of the nation's system of communications for what it sees as the benefit of the nation. The number of vehicles in the UK carrying goods and passengers increases year by year, creating a demand for better roads. When it comes to local level the space for these roads is not available. Rarely has a corridor been left for a new road or extra space alongside an existing road. To widen or provide a road in a town invariably involves some demolition of buildings.

(b) *Planning blight isolating a building*
During periods of urban renewal of obsolete dwellings there may be some buildings amongst them which are still viable. Where a commercial developer requires a specific area of land for redevelopment these would be obstructions to a comprehensive redevelopment. Where the local authority is undertaking the development it may acquire the buildings under a compulsory purchase order to allow them to be demolished and a new environment created. A private developer can only try to induce the 'obstructing' building owners to sell their property. It has been

known for owners not to sell and to delay development for many years.

It may be that the owner of the land around the affected buildings will demolish the adjoining property and, literally, leave the building isolated. This could leave it exposed and lead to an acceleration in its process of deterioration.

The area in which the building is located may undergo a social change, so that the building is isolated in 'use' terms. For example, an area with housing, shops and small factories might be demolished, leaving one or two factories, but the new development could be solely housing.

(c) *Value of land exceeds value of building*
In one of the above situations it may be that the value of the land rises to a level which is much higher than that of the building itself. The building may still be economic to run and have many years of structural life, but because of a change in status of the land upon which it stands it will be worthwhile for the owner to sell, demolish or redevelop for himself. For example, if new technologies are centred in new factories in a particular area, owing to ideal infrastructure services and environment, the demand for these will inflate and the land value in the locality will rise, which making the building plot more valuable.

Renewal and recycling are constant. From time to time there may be a bias towards recycling as opposed to new build. This could be due to government economic policies encouraging recycling by the giving of grants. Or, concern may be expressed regarding the wholesale movement of people away from a housing area into new dwellings. By maintaining the existing properties the community's social fabric will not be destroyed. According to some commentators, a contributory factor to the social problems on some local authority housing estates is the uprooting of people from one environment to another. They do not go en masse to the new estate, so a completely new mix of people is created. In future, when it becomes necessary to demolish estates and environments in a wholesale manner, greater regard will need to be given to the social aspects.

From the point of view of people in the construction industry, earning their living from the process and the product, it could be said that it is in their own interests for buildings to have a relatively short life cycle so that there is a continuous demand for new work and rebuild. Obviously, this is an extremely narrow view to take, which discounts the needs and wishes of the clients. Clients require buildings which are fit for purpose, inexpensive to run and will last as long as required. But there comes a point where the construction industry should advise the client that it is better to undertake some form of recycling. At what point in time that is reached will depend on the contextual framework factors described earlier. The decision to undertake, adopt and execute a technological decision on the recycling of buildings comes from an analysis and evaluation of these factors. We have therefore come full circle, in that to continue the development of the built environment we must seek its rationale within the issues, criteria, facets and conflicts emanating from the contextuial framework explored at the beginning of this book. The end of one building typically makes the beginning of another.

SUMMARY

Initially some definitions of rehabilitation, adaptation/refurbishment, and preservation/conservation were considered. It was seen that there were overlaps between these terms in practice. There were eight factors which could influence the eventual decision relating to the recycling, or otherwise, of the building. At any one time any of these or a combination of some or all could lead to a particular decision. A brief comment was made with respect to energy in a building's recycling. A list of construction constraints was given which indicated some of the problems which could affect the technologies of site work. Finally the relative transience of buildings was discussed and it was recognised that sooner or later most buildings will be demolished to make way for new. The conclusion thus referred back to the first section of the book which described the contextual factors determining the building technology.

QUESTIONS

1. Describe the building technologies appropriate to the following: rehabilitation; refurbishment; conservation. Consider the process from design to production.
2. Discuss how the nature of the construction industry could determine the course of action in recycling a building.
3. To what extent must the construction constraints be considered in the recycling of a building?
4. Should we discount the energy lost when demolishing a building and put this down as a price to pay for progress?
5. Taking one of the contextual framework factors show how this can influence a recycling decision.

Further Reading

ALEXANDER, C. *The production of houses*, Oxford University Press, 1985.

ASTM/CIB/RILEM, *Performance concept in building* (Proceedings of the 3rd ASTM/CIB/RILEM Symposium), 1982.

BAMPTON, E., 'Presenting the owner's case', in *Building better buildings – maintenance at the design stage*, Chartered Institute of Building Maintenance Information Service No. 3. Spring 1978.

BARTLETT INTERNATIONAL SUMMER SCHOOL, *Production of the Built Environment*, University of London, 1982.

BENDER, A., *A crack in the rear view mirror*, Van Nostrand Reinhold, 1973.

BENSON, J., EVANS, P., COLUMB, P. AND JONES, G., *The housing rehabilitation handbook*, Architectural Press, 1980.

BENTLEY, M.J.C., *Quality control on building sites*, Building Research Establishment. CP 7/81.

BERNSTEIN, D. AND STASIOWSKI, F., *Project management for the design professional*, Architectural Press, 1982.

BOWLEY, M., *The British building industry*, Duckworth, 1966.

BRAGG, S.L., *Final Report of the Advisory Committee on Falsework*, Health and Safety Executive. HMSO. 1975.

BRANDON, P., *Building cost techniques: new directions*, E. & F.N. Spon, 1982.

BRANDON, P. AND POWELL, A., *Quality and profit in building design*, E. & F.N. Spon, 1984.

BRANDON, P.S., MOORE, R.G. AND MAIN, P., *Computer programs for building cost appraisal*, Collins 1985.

BRITISH PROPERTY FEDERATION, *Manual of the BPF System: The British Property Federation System for Building Design and Construction*, BPF, 1983.

BRITTEN, J.R., *What is a satisfactory house? A report of some householders' views*, Building Research Establishment, Current Paper 26/77.

CIRIA, *Buildability: an assessment*, Construction Industry Research and Information Association Special Paper 26, 1983.

COOPER, D.J., 'Maintenance management today and tomorrow – a manager's view', in *Building maintenance – present and future*, Chartered Institute of Building Maintenance Information Service No. 1, Autumn 1977.

COVINGTON, S.A., *The degree of quality assurance provided with certain building components and products*, Building Research Establishment Current Paper 8/80.

FISK, D.J., *Microeconomics and the demand for space heating*, Building Research Establishment Current Paper 6/78.

FORBES, W.S., *The rationalisation of house building*, Building Research Establishment Current Paper 48/77.

FLANAGAN, R. AND NORMAN, G., *Life cycle costing for construction*, STS Publications, 1983.

FREEMAN, I.L., *Comparative studies of the construction industries in Great Britain and North America: A review*, Building Research Establishment Current Paper 5/81.

GARDNER, E.M. AND SMITH, M.A., *Energy costs of house construction*, Building Research Establishment Current Paper 47/76.

GRAY, C., *Buildability – the construction contribution*, Chartered Institute of Building Occasional Paper No. 29, 1983.

HARPER, D.R., *Building: the process and the product*, Construction Press, 1978.

HERBERT, G., *Pioneers of prefabrication*, The Johns Hopkins University Press, 1978.

HILLEBRANDT, P., *Analysis of the British building industry*, Macmillan, 1984.

HUTTON, G.H. AND DEVONALD, A.D.G., *Value in building*, Applied Science Publishers, 1973.

HIRSCH, F., *The social limits to growth*, Routledge & Kegan Paul, 1977.

LANDES, D.S., *The unbound Prometheus*, Cambridge University Press, 1969.

LAMSAC, *Terotechnology and the maintenance of local authority buildings*, Local Authorities Management Services and Computer Committee, 1981.

MARTIN, B., *Standards and building*, Royal Institute of British Architects, RIBA Publications Ltd, 1971.

MAINTENANCE PRACTICE COMMITTEE, *A guide to maintenance management*, Chartered Institute of Building.

MARKUS, T.A. (ed.), *Building conversion and rehabilitation. Designing for change in building use*, Newnes-Butterworths, 1979.

MILLS, E.D. (ed.), *Building maintenance and preservation. A guide to design and management*.

NATIONAL BUILDING AGENCY, *Maintenance procedures for housing associations*, National Building Agency, 1977.

NEDO, *Cosntruction into the early 1980s*, National Economic Development Council, 1976.

NEDO, *Competence and education*, National Economic Development Council, 1984.

NEDO, *Faster building for industry*, National Economic Development Council, 1983.

NOBLE, V., *The value of building maintenance*, Chartered Institute of Building Maintenance Information Service Paper No. 13, 1980.

PAWLEY, M., *Garbage housing*, Architectural Press, 1975.

PETTIT, R., *Computer aids to housing maintenance management*, Chartered Institute of Building Management Information Service No. 10, 1979.

RUSSELL, B., *Building systems, industrialisation and architecture*, John Wiley, 1981.

SCOTT, G., *Building disasters and failures*, Construction Press, 1976.

SCHUMACHER, E.H., *Small is beautiful*, Blond and Briggs, 1973.

SEELEY, I.H., *Building maintenance*, Macmillan, 1982.

SKINNER, N.P., *Interpreting feedback information – some examples from housing maintenance*, Building Research Establishment Current Paper 1/83.

SKINNER, N.P. AND KROLL, M.E., *Maintenance feedback*, Chartered Institute of Building Maintenance Information Service No 18, 1981.

SULLIVAN, B.S., *Industrialisation in the building industry*, Van Nostrand Reinhold, 1980.

TAVISTOCK INSTITUTE, *Interdependence and uncertainty*, Tavistock Institute, 1963.

TURIN, D.A., *Aspects of the economics of construction*, George Godwin, 1975.

WHITE, R.B., *Prefabrication – a history of its development in Great Britain*, HMSO, 1965.

Index